PSYCHOPHARMACOLOGY
A STEP-BY-STEP PROGRAMMED GUIDE

Perry W. Buffington, Ph.D.
M.S. Clinical Psychopharmacology

This book is for educational purposes only. It does not purport to diagnose, treat, or offer any medical advice. Nothing takes the place of a licensed healthcare provider/professional.

Special Thanks to:

Miranda Barr, proofreading, editing, and researching

Misty P. Smith & William L. Taylor, charts

Printed in the United States of America

ISBN 978-1-7359340-3-7

Dedication

For all my students, each one.

About the Author

Perry W. Buffington, Ph.D., M.S., initially trained as a psychologist, has taught at the graduate and undergraduate level for over thirty years; served as contributing editor for Delta Air Lines' in-flight magazine, *Sky*, writing over 120 consecutive monthly articles; hosted his own syndicated radio show through Envision Networks; produced and appeared in weekly television segments in the Atlanta and Jacksonville (FL) markets; served as a syndicated newspaper columnist for Universal Press; and authored over 13 books, including his best-selling series, *Cheap Psychological Tricks* (Peachtree Publishing, Atlanta). Returning to school, he was graduated with an M.S. degree in clinical psychopharmacology and passed the "Psychopharmacology Examination for Psychologists (PEP)." He is known for his highly popular continuing education seminars in psychopharmacology, having taught hundreds of them. He currently serves as a faculty member, Psychology Department, University of Georgia. This step-by-step programmed guide is his first textbook and follows his lectures from his classroom psychopharmacology courses.

Table of Contents

Tables and Charts

INTRODUCTION

Not so long ago, a student entered my graduate psychopharmacology class, promptly took a seat up front, raised his hand, and said, "You know, Professor, we are all scared of this class." The class grew quiet, waiting for a response. He expressed what every student was thinking.

Hardly the first time a student (or modern practitioner in a continuing education class) has expressed this concern, students may not say it out loud, but a good professor knows. Their faces say it all. Students walk in, take their seats, fear all over their faces, expressing anxiety that no benzodiazepine will fix. Why? Pharmacology (and its baby brother, "psychopharmacology") is known throughout all university classes – Ph.D. and master's programs, medical school, nursing programs, physician assistant programs – as tricky, hard to master, and the bane of existence for most students.

Back to the openly trepidatious student: Acknowledging his fear, worry, anxiety, "You are right; it is a tough class, but there is a way to stack the dominos."

And that is what this programmed text does best. It stacks the information so that one domino of knowledge builds on the next domino, which serves as a prerequisite toward understanding the following. Before you know it, the tiles of drug knowledge are arranged in a way that a method to the madness is revealed.

First, a broad foundation is required. Once the basic vocabulary is in place, it is time to understand more esoteric aspects of pharmacology (e.g., pharmacokinetics and pharmacodynamics, agonist vs antagonist). Now add to that a cursory understanding of neuronal functioning, and it is time to learn receptors. After mastering these foundational layers, it is all about medications.

Psychopharmacology is like learning a new language - except that the words make no sense. In some foreign languages, the terms "sound like" something you have heard before. However, "psychopharm-ese" is full of words like phenylethanolamine N-methyltransferase. And to complicate matters, psychopharmacologists like to abbreviate these big words; in this case, this big word is abbreviated "PNMT." Students come into psychopharmacology class afraid of its difficulty, syntax, terms, body of knowledge, and complicated amalgamations. Yet a basic understanding of psychopharmacology is essential in today's integrated care. It all starts with a foundation, a firm footing, upon which one can build a strong knowledge base.

This book is designed for university students and professional "students" — anyone who wishes a firm grasp on both the "science" and "art" of psychopharmacology. Even if you are a seasoned prescriber, you are still a student. To that point, everyone in this field is a student and remains so for their life. In psychopharmacology, continually learning is required. New studies, new findings, new guidelines often emerge, frequently, and regularly. Moreover, now that the field of psychopharmacology is morphing into the new area of molecular biology/pharmacology, everything is up for review. Bottom line, in this field, everyone must remain professional students.

What is the first step in this programmed learning approach? Follow the directions listed in the following "HOW TO USE THIS PROGRAMMED GUIDE." Upon successful completion of all fifty modules, one will have (1) a dramatically increased understanding of the field of psychopharmacology, (2) a psychopharmacology vocabulary that improves one's communication with other mental health providers, and (3) a more straight-forward approach toward protecting patients. This programmed textbook represents fifty initial steps toward the mastery of this complicated science.

This book is unique because it is geared more toward mastery than acquiring a "useless" body of knowledge. Instead, this is a step-by-step, each module builds upon the next, programmed text. That means the body of essential information (defined as topics that are included in most university textbooks) has been broken down into manageable, digestible bites. Some are easy; some are hard, but each case is a precursor, a prerequisite toward better understanding and appreciating the next topic.

With that in mind, following these guidelines will allow you to maximize learning and to reduce the "We are so scared about this course!" feeling. Here is how to tile the dominos, stack the blocks, build a psychopharmacology house, and start mastering the field.

(1) Start with Module 1. Sounds silly but start at the beginning. Regardless of your level of knowledge and to get the most out of this book, start with Module 1 and work your way forward. After each module, complete the self-assessment quiz at the end. Achieve at least 80% (four out of five questions – some modules have more questions) correct answers, move on. Frankly, it is better if you achieve greater than 80%; that makes the next module easier.

(2) Do not skip a module. Start with one, then move on to two, continue with three, and so on. Skipping modules makes it difficult for the knowledge to consolidate. Do not be alarmed if this approach to studying psychopharmacology feels

disjointed. It may feel like a mixture of unrelated things strung together without a common theme. With consistent, guided study, the material will amalgamate into a coherent "I get it."

(3) Before starting a new module, look over the previous module's questions and answers. It will take less than a minute. This review serves as a stimulus reinforcer and continues to keep these concepts forefront.

(4) Consider flashcards. See something interesting, jot it down on a notecard. Walking, sitting in a boring meeting or resting at the end of the day, reviewing them keeps these concepts top of mind.

(5) Reread and review. If there is an area you do not understand, go back, reread, and review until you can correctly answer 80% of the questions at the end of the module.

(6) Note the "smiley faces" [☺] at the beginning of each module. They provide a clue as to the difficulty level associated with that section.

☺ = Difficult, probably need to study and reread this section.

☺☺ = Hard, complex, complicated. Requires thinking.

☺☺☺ = Not too hard; not too easy.

☺☺☺☺ = Easy, but usually a twist that is slightly complicated.

☺☺☺☺☺ = Easy. A fun read.

Even if you are a seasoned prescriber, if you are not a prescriber, or if you never want to prescribe, as you read this book, you will understand how these drugs do what they do. You will get a handle on why the drugs do not do what you want them to do. You will have a better understanding of side effects. All this combined make for a better clinician, a better advocate for your patients, and a better consumer of medications for yourself. It is all about how drugs work and perhaps more importantly, why they do not work.

It is time to take the dominos out of the box and start arranging the tiles in a coherent pattern. Start with Module 1, "A Drug by Many Other Names – Psychotropics." It is an easy, five "smiley faces" straight-forward module. Only a few pages long, it can be quickly completed in one sitting. Think of it as a first lesson in the foreign language called "Psychopharmacology."

That said, drug names certainly sound like a foreign language and can be both off-putting and daunting. Students struggle to pronounce drug names like venlafaxine, chlordiazepoxide, desvenlafaxine, even age-old chlorpromazine. Remember this rule, one that comes to the rescue when you see a new generic drug name: <u>Pronounce each syllable equally.</u>

Venlafaxine becomes "ven-la-fax-ine." Desvenlafaxine? Same as venlafaxine, just put a "des" in front of it. Chlorpromazine: Chlor-pro-ma-zine. Do not worry about inflection or accents. Just pronounce each syllable equally. It just takes practice and a little familiarity.

It is an art of no little importance to administer medicines properly, but it is an art of much greater and more difficult acquisition to know when to suspend or altogether to omit them.

Philippe Pinel, 1809

Treatise on Mental Alienation

A Drug by Many Other Names

Psychotropics

☺ ☺ ☺ ☺ ☺

William Shakespeare was wrong. To paraphrase from *Romeo and Juliet* (1597), this is one time when a name by any other name is nowhere near as sweet or as correct. The class names for drugs in the field of psychopharmacology are tricky. There are a lot of the esoteric terms and nomenclature which describe psychopharmacological drugs. They sort of look the same; they do not mean the same. They seem simple enough, even innocuous, and many believe they can be used interchangeably; but they are subtly different. If a clinician calls these drugs or drug class by the wrong name, they unwittingly open one's abilities to criticism. So, step one is to understand the subtle differences in the overall category names that refer to the drugs used in this field. Fortunately, there is one term when used, no matter what class of drug is being described, that will always be correct.

First off, start with the word used to describe the field. Psychopharmacology. It is a subset of the field of pharmacology. It technically may be the youngest, even though many of the medications in psychopharmacology have been used as therapeutic herbs, "worts," and flowers for thousands of years without knowing exactly how they worked.

The term, "psychopharmacology," was first used in the 1920s but did not enter the dictionary until the 1950s – one can thank the creation of Thorazine®/chlorpromazine for that. Psychopharmacology is the study of medications that affect the mind in the treatment of psychiatric or psychological disorders. Within psychopharmacology, there are class names used to describe drugs under the psychopharmacology umbrella (e.g., antidepressant, antipsychotic, mood stabilizers, and others).

Take a look at the four terms, which are thought to be synonymous with psychopharmacology agents. The four terms that broadly define all drugs used in psychopharmacology are psychoaffective, psychotherapeutic, psychoactive, and psychotropic. One may think they mean the same, but there are some differences. Of this group, there is one term that is always correct – no exception. That may be the only time the two words, "no exception," are used in this entire, programmed textbook.

(1) Psychoaffective. This is a psychopharmacological agent that affects "mood." The word "affect," in psychological parlance, is a synonym for mood. Putting it together, a psychoaffective is a "mind-mood agent." This term is used, but it is likely more pedantic than the other three. Typically, these medications are referred to as "psychoaffective agents."

(2) Psychotherapeutic. Easy one. It is a pharmacological agent (aka "drug") that is used "therapeutically" to treat problems associated with thought processes, perceptual distortions, and behavioral disorders. Professionals use this word, but part

of the problem with this word is that the drugs may not be as "therapeutic" as one would like — too many side effects, does not work at all, intolerance, too expensive, the list goes on. The term psychotherapeutic may provide false hope. The next term, psychoactive is the one that trips most people up.

(3) Psychoactive. This one is tricky. While many think these four terms are interchangeable, this term is the one term that is not. There are two rules to remember with psychoactives. First, they are drugs that, by definition, work within the Central Nervous System (CNS, brain and spinal cord). Second, they typically work quickly, in as little as twenty minutes. All psychoactives are psychoaffectives, psychotherapeutics, and psychotropics; however, not all psychotropics, psychotherapeutics, and psychoaffectives are psychoactives. Remember, psychoactives must work quickly in the CNS.

Prozac®/fluoxetine is a psychotropic, psychoaffective, and psychotherapeutic. Prozac®/fluoxetine does **NOT** work in 30 minutes in the central nervous system (CNS), so it is **NOT** a psychoactive.

Valium®/diazepam kicks in quickly – thirty or so minutes. It works in the CNS. Therefore, it is a psychoactive – a psychotropic, psychoaffective, and psychotherapeutic, too. If that is a little confusing, not to worry. The winning term, psychotropic, will resolve any uncertainty.

(4) Psychotropic. The winning word. This is the word to use when one does not know which one of the four terms is correct. Coined in the late 1950s, take a look at this word. Break it down. The "psycho" in psychotropic means "mind." "Tropic" from

the Greek word, "trepein," means "to turn." Put the two together, and a psychotropic is a word that means "to turn the mind," and that is exactly what it does. The U.S. National Library of Medicine classifies psychotropics as a "loosely defined" grouping of drugs that have effects on psychological function. Including, but not limited to antidepressants, hallucinogens, tranquilizing agents, and antipsychotics.

Here is the big point, the takeaway: If one does not know what term to call these pharmacological agents, if one does not know precisely in which class a particular psychopharmacologic drug fits, and if one is unsure whether it is a psychoactive, then use the term psychotropic. Take it a step further, calling all psychopharmacological drugs psychotropics is always correct. There is no exception to this rule.

QUIZ TIME

(1) Psychopharmacology is a subset of the broader field of _____.

(2) The "affective" part of psychoaffective means _____.

(3) From the list of four terms, _____ and _____ may be the most interchangeable.

(4) Of the four terms, all _____ are psychotropics, but not all psychotropics are _____.

(5) If you are not sure which term to use to describe a specific medication within psychopharmacology, use _____, and you will always be correct.

ANSWERS

(1) pharmacology

(2) emotion or mood

(3) psychotherapeutic, psychoaffective, and psychotropic, too. Any two of these three are correct.

(4) psychoactives; psychoactive

(5) psychotropic

Score three or less, probably need to reread the module.

Score four or more, move on to the next module.

All five correct? Gold star. Well done. On to bigger and better molecules.

Module 2

The Name Game

Generic vs Brand

☺ ☺ ☺ ☺ ☺

One common theme through this field of study is that there are always several names, synonyms for the same thing. Later in this programmed book, it becomes obvious that everything in pharmacology and psychopharmacology is expressed backward, sort of a negative feedback loop. Right now, start with individual drug names.

Each drug has not one, not two, not three, but four names for the same thing. Each medication has chemical, generic, brand, and street names. The first term is the one rarely discussed and likely never used by professionals. To explain, take a look at all of Valium®/diazepine's names, in the order the names were given.

(1) <u>The Chemical Name</u>. This is the chemical description for synthesizing the drug utilized by organic chemists. The chemical name of Valium®/diazepam is written as follows: 7-chloro-1,3-dihydro-1-methyl-5-phenyl-2H-1,4-benzodiazepin-2-one. The chemical formula defines the drug, describes its structure and molecular targets, and follows the guidelines delineated by the International Union of Pure and Applied Chemistry (IUPAC). For most people, that is just chemical "mumbo jumbo." Unless fluent with a mortar and pestle and desire to whip up a batch in your home laboratory, all one needs to know is that all drugs have a chemical name, a laboratory

chemist's recipe. In this case, the chemical name is the precursor and two-steps away before the drug "Valium®" is given a "proper marketing name."

(2) <u>The Generic Name</u>. This name always follows the chemical name. In the case of "Valium®," it is "diazepam." Drugmakers create generic names according to the chemical makeup of the compound. Generic drugs/names are not protected by trademark. So, "diazepam" is the generic form of the proprietary medication, "Valium®." Note, that "diazepam" is not capitalized. Only the brand medication is capitalized.

For the sake of trivia, "generic" comes from Latin, "gene" or "gener" and means "give birth," which is the stem of the scientific word, "genus." It was first used to describe medications in 1953. In terms of products, which can also be related to medications, it means "not special," "not brand name," and was originally meant to signify something "cheap." However, that was then, and this is now. Generic medications can be pricey (Eban, 2019).

If the drug is marketed in the U.S., the United States Adopted Name Council (USAN) assigns the active ingredient of the drug a generic name. The generic names are official, non-trademarked names, reviewed and cleared by the International Non-Proprietary Name (INN) program, and managed by the World Health Organization (WHO). The generic name, while often difficult to pronounce, especially for patients, is intended to provide a unique standard name for each active ingredient. Most people are not familiar with the INN system, even though the WHO has coordinated it since

1953 – remember, that is the year the word "generic" appeared for the first time (Karet, 2019).

As one grows more familiar with generic names, notice that the "ending suffix" will usually provide or at least suggest a clue as to the class of drug or the "why" this drug is being prescribed. In the case of "diazepam," the prefix is "pam." Any medication that ends in "pam" or "lam" (usually, but not always, there are exceptions to <u>everything</u> in psychopharmacology) offers a clue that this drug is a benzodiazepine – usually an antianxiety (or anxiolytic) medication.

For example, generic drugs that end in "avir" or "ivir" are usually antivirals. If it ends in "ole," think antiparasitic. "Caine," local pain relievers. Or "statin," anti-hyperlipidemic. Between the IUPAC, the USAN Council, the INN program, and the WHO, there is a method to the naming madness.

<u>(3) The Brand Name.</u> This is the name of the drug that is usually heard first, especially on all the television commercials; but interestingly, it is usually the third name given to a drug (and certainly the most creative). If it is popular, this name will stick. However, before "branding," the two previous names take priority.

The brand name, "Valium®," is always capitalized. Think of it as its unique advertising, marketing, brought-to-market name. Because it is difficult to register formally and legally (to make sure that it is not used elsewhere), it is the hardest to decipher. Of course, there is a governmental agency involved: The FDA's Division of Medication Error Prevention and Analysis (DMEPA). Additionally, two more departments within the FDA scrutinize brand names: The Center for Drug Evaluation and Research (CDER) and the Center for Biologics Evaluation and Research (CBER). They reject approximately 30% of submitted names. Additionally, when the drug goes global, there are many other governmental agencies involved – all designed for consumer safety.

Think of brand names as creative marketing, creative engineering, and/or creative subliminals. How brand names are selected is a deep, dark secret. Millions of dollars are doled out to creative advertising firms just to give the brand name, proprietary drug, the "perfect" name – one that will capture the public's attention, will be easy to advertise, and will unwittingly persuade the consumer that "this is the perfect drug for you, the community, the state, the nation, and the world."

Usually, these names are catchy, have a "z" or an "x" somewhere in their name, and are designed to "subliminally" provide some modicum of assurance that this drug does what it is supposed to do.

THE NAME GAME

Drug Name	Derivation
Actigall®	**ACT**s on the **GALL** bladder.
ursodiol (generic of Actigall®)	Ursus is Latin for "bear." Originally discovered in bear bile. For gallstone dissolution and prevention
Batroban®	**BAN** bacteria (Ba)
Glucophage®	**GLUCO** is from "glucose," and **PHAGE** is from Latin, "to eat." EATS GLUCOSE, treats diabetes
Singul**air**®	For breathing, Singul**air**® (see "AIR" in the name)
montelukast (generic of Singulair®)	montelukast is the generic of Singulair®. The generic name identifies the drug's hometown, Montreal
Tylenol®	Derived from chemical name of compound, N-ace**TYL**-para-aminoph**ENOL** (APAP)

(4) The Street Names. This is the last of the names "given" to any medication. The more abusable the drug, the greater number of "street names." They are slang terms given to common medications – often used to disguise what the abuser is talking about from the uninitiated. New street names are added frequently. For instance, the common street names for Valium® are "Vs," "Yellow V's" (named for the 5 mg tablet), "Blue V's" (named for the 10 mg tablet), "Benzos" (shortened version of the class of medications, benzodiazepines), "Dead flower powers" (named in honor of the many overdoses in the 1960s), "Downers," "Foofoo" (named for the fancy-frilly

feelings the drug provides), "Howards" (named in honor of Howard Hughes, a notorious user), "Tranks" (a shortened version of "tranquilizer"), "Drunk Pills," and many, many more. Because these names vary by region, the force of law enforcement, dosages, age brackets of users and sellers, and phenomenon associated with their use, good luck keeping up with the constantly changing names. For Prozac®/fluoxetine's multiple names, check out the following table.

PROZAC® MULTIPLE DRUG NAMES	
Chemical Name	N-methyl-3-phenyl-3-[4-(trifluoromethyl)phenoxy]propan-1-amine
generic (no caps)	fluoxetine
Brand	Prozac® — allegedly named for prefix "pro," meaning positive, and "zac" sounds like "zip" or "zap." Subliminal translation/marketing spin: "positive zap" to your depression
Street	Not an abusable drug; little street value, few street names available. Originally, Prozac® was referred to as "Bottled Sunshine," "Wonder Drugs," "Miracle Pills," "Happy Pills," and "Bottled Smiles"

QUIZ TIME

(1) Rank the order from 1 to 4 (with 1 being first, and 4 being last) in which drugs receive their names.

     _____Generic

     _____Brand

     _____Street

     _____Chemical

(2) True or False? The prefix (beginning) of the generic medication's name suggests the drug class to which this generic belongs.

(3) True or False? A generic drug's name is not capitalized.

(4) True or False? It is essential that all clinicians know both the generic and brand name of any medication.

(5) Of the four terms (psychoaffective, psychotherapeutic, psychoactive, and psychotropic), all _____ are psychotropics, but not all psychotropics are _____.

ANSWERS

(1) Drugs are named in chronological order: (1) chemical, (2) generic, (3) brand, (4) street. Of the four, street names are ever-changing; but chemical, generic, and brand remain constant.

(2) FALSE. It is the suffix of the medication that gives a clue to its drug class.

(3) TRUE. Only the brand name is capitalized.

(4) TRUE. Students of psychopharmacology always want to know: "Do I need to know both generic and brand names?" The answer, "Unequivocally, yes." Often patients will use one or the other, so you must be mindful of both. Also, knowing the name of both generic and brand reduces the probability of a script-writing drug mistake.

(5) All psychoactives are psychotropics, but not all psychotropics are psychoactive.

FOR THE FUN OF IT. . . .In addition to chemical, brand, and generic names, often, through everyday use, pharmaceuticals earn funny nicknames – slang names that are given by patients, prescribers, emergency room staff, even drug sales staff. Look at the following and match the nickname to the drug. A drug can have more than one slang nickname.

_____1. "No Effexor" A. Prozac®/fluoxetine

_____2. "Pfizer Riser" B. Haldol®/haloperidol

_____3. "Dopa-Max" C. Effexor®/venlafaxine

_____4. "Side Effexor" D. Ativan®/lorazepam

_____5. "Vitamin H" E. Topamax®/topiramate

_____6. "Vitamin A" F. Viagra®/sildenafil

_____7. "Bottled Sunshine" G. Ritalin®/methylphenidate

_____8. "Vitamin R"

"FOR THE FUN OF IT" ANSWERS

1. C, "No Effexor"/Effexor®/venlafaxine. Nicknamed after its brand name, Effexor®.

2. F, "Pfizer Riser"/Viagra®/sildenafil. Nicknamed after the company that created it (Pfizer), and for its common use in the real world (riser).

3. E, "Dopa-Max"/Topamax®/topiramate. Nicknamed for its cognitive dulling ("Dopa-Max") and "tip-of-the-tongue" side effect, anomia.

4. C, "Side Effexor"/Venlafaxine's brand name, Effexor® earns a double nickname moniker, both "No Effexor" and "Side Effexor."

5. B, "Vitamin H"/Haldol®/haloperidol. Named for its neuroleptic (original meaning: "nerve seizing") action for agitated patients who, in acute need, arrive at the hospital's emergency department.

6. D, "Vitamin A"/Ativan®/lorazepam. The counterpart of "Vitamin H." Named for its calming influence for agitated patients who, in acute need, arrive at the emergency room.

7. A, "Bottled Sunshine"/Prozac®/fluoxetine. How did it get this nickname? Marketing.

8. G, "Vitamin R"/Ritalin®/methylphenidate

Say My DNA Name

Pharmacogenomics

It is time to look at two significant terms plus one more that represents the future of psychotropic medications. The first two terms are pharmacokinetics and pharmacodynamics. More, lots more, on those later. The additional term to know, pharmacogenomics (aka "pharmacogenetics"), is the future of this field. It will clearly play a remarkable part; but a basic understanding of pharmacogenomics will be sufficient for right now. Imagine the following patient-prescriber scenario – the usual drill, business as usual, with one added step.

Feeling under the weather?

Step 1: Head to your prescriber's office.

Step 2: They look you over.

Step 3: They check your vitals.

Step 4: They evaluate your symptoms.

In the past, the prescriber would have jumped, prescribed treatment, and written a medication script (or two). However, with pharmacogenomics, there is a new, added step.

Step 5: Before prescribing a medication, the prescriber tests your DNA (either by cheek swab, drop of blood, or blood draw) and waits for the results.

Step 6: Time to prescribe a "smart drug." This is a customized treatment, a medication tailored to your genes, prescribed just for you. While the current standard of care is to specify an average dose based on one's body size and age, tomorrow's treatments will be based on the body's needs and centered on a patient's specific genomic variations based on what makes an individual unique.

Pharmacogenomics (referred to as "PGx"), how an individual's genes influence their specific response to drugs, is a relatively new, emerging field that combines the science of drugs and the study of gene function. The goal is to develop safe and efficacious medications tailored to a person's genetic makeup by combining these two. As a result, "personalized medicine or prescribing" is often used to describe pharmacogenomics.

Much of today's psychotropic prescribing, by necessity, is more like a "shotgun" or a "one-size-fits-all" approach. The result is a drug that works but may or may not work the same way for every person who takes the med. For example, for over fifty years, antidepressants have been prescribed without specific genomic rubrics, relying instead on the "art of prescribing," tantamount to "trial and error." With the current "shotgun" approach, it is difficult to predict how a patient will respond and its problematic side effects. As a result of the Human Genome Project (which identified over 21,000 genes), drug manufacturers and researchers study how inherited differences in genes determine the individual's drug response. The goal is to finally develop a "magic bullet" first discussed by Paul Ehrlich in the late 19th Century. A "Magic Bullet" or "smart drug" is a medication that heals without causing harm.

But remember: At present, this testing cannot guarantee the best response to a medication, but it can help a provider predict whether a patient is likely to experience unwanted side effects. The cost is usually a few hundred dollars; some insurers and government agencies may reimburse. While pharmacogenomics is still taking baby steps, this present and future field offers great promise.

To that point, the Department of United States Veterans Affairs is conducting the PRIME Care Study (Precision Medicine in Mental Health Care). From their website, they explain:

> The purpose of this study is to determine the effectiveness of providing depressed Veterans and their providers the results of pharmacogenetic (PGx) testing for psychotropic medications. The study focuses on whether and how patients and providers use genetic test results given to them at the time an antidepressant is to be initiated to treat Major Depressive Disorder (MDD) and whether use of the test results improves patient outcomes. MDD is one of the most common conditions associated with military service and combat exposure, increases suicide risk, and worsens the course of common medical conditions, making it a leading cause of disability and mortality. Validation of a PGx test to personalize the treatment of MDD represents an important opportunity to improve the healthcare of Veterans. Veteran recruitment will take place at many VA sites (www.bit.ly/VAPRIMECare).

One final point on pharmacogenomics/pharmacogenetics, while this field of study holds great promise, currently there are at at least two risks. First, there is always the risk that one's genetic information can be used as a future identifier in some George Orwell *1984* dystopian scenario. Second, there is the risk that one's genetic information may be used against an individual. To explain, in 2008, the Genetic Information Non-Discrimination Act (GINA) was passed into law. This law does in fact protect against discrimination by one's health insurer and employer; however, it does NOT protect against discrimination by life insurers and long-term care insurers.

The goal of PGx is better care and reduced costs. It is the future, and it is here.

QUIZ TIME

(1) True or False? Pharmacogenetics and pharmacogenomics are synonymous terms.

(2) True or False? All drugs have four names: professional, brand, generic, and street.

(3) True or False? The goal of pharmacogenomics is to find a "magic bullet" that goes directly to the diseased site, bypasses healthy tissue, does its job, and then disappears, causing no harm to the individual.

(4) True or False? Presently genetic testing is in its infancy, emerging benefits and risks are not completely understood, and patient protection is evolving with the field.

(5) True or False? Pharmacogenomic testing cannot guarantee the best response to a medication, but it can help predict whether you are likely to experience unwanted side effects.

ANSWERS

(1) TRUE. These terms are used interchangeably, and both are abbreviated "PGx."

Admittedly "pharmacogenomics" sounds more scholarly than "pharmacogenetics," but they are the same.

(2) FALSE. "Professional" is the person who prescribes the drug. The four names are chemical, brand, generic, and street.

(3) TRUE. That is the goal. That is the hope. One day it will be achieved. Smart drugs will be the norm. The positive result will be a treatment that works the first time with fewer side effects, less medical errors, and minimal drug mistakes.

(4) TRUE. Benefits and risks are evolving together, but how to reduce risks – an ethical and governmental responsibility – may be lagging. This is likely since the field is proliferating so rapidly that society cannot keep up. Risks and benefits are yet to be determined or the need to protect may not be a priority.

(5) TRUE. Currently, genomic testing is a good start. It is a beginning point to help the prescriber utilize their knowledge base combined with test results to determine the best medication for the patient. However, solely relying on genomic testing could easily become "prescribing for dummies." If the clinician relies solely on these tests, without an exacting knowledge of how the drug works or what the drug does, patient safety will suffer.

Method to the Madness

Mechanism of Action (MOA)

☺ ☺ ☺ ☺

Pharmacokinetics and pharmacodynamics are coming up; but first, understanding the expression, "Mechanism of Action (MOA)," will make these two terms easier to process. The first two rules students learn in pharmacology and psychopharmacology are (1) all drugs have a mechanism of action, and (2) all drugs have side effects.

First, all drugs have a mechanism of action (MOA). All drugs do something specific at a place of action (e.g., receptor). For instance, if asked what is Prozac®/fluoxetine's MOA? For one of the first and probably most well-known antidepressants, the answer is SSRI. What makes this complicated is that the drug's mechanism of action is always abbreviated. At this time, SSRI's exact meaning may be unfamiliar. Break it down: SSRI stands for "Selective Serotonin Reuptake Inhibitor." There is an upcoming module that will explain "selective" and "reuptake inhibition."

Try another, take Risperdal®/risperidone; its mechanism of action is "SDA" and that translates to "Serotonin Dopamine Antagonist." The "A" in a drug's mechanism of action is always tricky. Will it be an agonist or an antagonist? Not to worry right now, that is coming up later as the foundation is built, term by term. Or how about

Remeron®/mirtazapine? It is a NaSSA – "Noradrenergic antagonist and Specific Serotonergic Antidepressant" and occasionally it is written "Norepinephrine Antagonist and Serotonin Specific Antagonist (or Agent)." Either way, it is a NaSSA. They all say the same thing. Right now, just remember one simple rule: All drugs have a mechanism of action. When taking a psychopharmacology course or entering clinical practice, these abbreviations are bandied freely as sort of a secret language between prescribers. Also, new terminology is being introduced, "Neuroscience-based Nomenclature (NbN)" and explained in upcoming modules. Whether one uses the old "MOA" terminology or the newer "Neuroscience-based Nomenclature (NbN)," there is one problem which can get the patient and the prescriber in unwanted side-effect trouble.

A drug's mechanism of action ONLY explains why the drug is being used therapeutically. It does not delineate everything the drug can do. In other words, the drug's therapeutic mechanism of action leaves out all the things a drug can do which may lead to therapeutic benefits, side effects, and/or adverse drug actions (ADR, defined as something very unusual and not often seen).

For example, when asked, "What is Elavil®/amitriptyline's mechanism of action?" the correct answer is, "It is an old TCA (**T**ri-**C**yclic **A**ntidepressant)." A better answer would be, "It is a SNRI," which stands for "Serotonin Norepinephrine Reuptake Inhibitor." The MOA tells a practitioner that the drug is being used therapeutically as an antidepressant. However, SNRI does not tell the entire story. It, the drug, is used for other things, too.

For years, this "TCA" drug has been referred to as a "dirty" drug or "pharmacologically rich" medication. This means it does more than wanted or required. Put another way, it activates more places (receptors) than a prescriber wants it to stimulate, and that causes unwanted side effects. For now, there is one rule with an inconvenient truth: All drugs have a mechanism of action (MOA), and the MOA does not state everything the drug does. It describes why it is being used therapeutically, but it leaves out a lot of information that can explain many side effects associated with this medication.

The second rule follows the first: All drugs have side effects. To that point, Elavil®/amitriptyline, in addition to helping relieve depression (that is the serotonin and norepinephrine part), activates other receptors not listed in the drug's mechanism of action and can cause (and this reads like a drug commercial): dizziness, dry mouth, urinary retention, constipation, tachycardia (rapid heartbeat), increased sedation, decreased sweating, confusion, visual problems, weight gain, and many, many others. Fortunately, and thankfully, all side effects do not happen with all patients, but there is always the possibility.

QUIZ TIME

(1) "MOA" stands for _____.

    (a) Modus Operandi Actum

    (b) Mechanism de Actio

    (c) Method of Action

    (d) Mechanism of Action

    (e) Mall of America

(2) True or False? All drugs have five names.

(3) True or False? All drugs have side effects.

(4) A "dirty" or "pharmacologically rich" drug is one that _____.

    (a) easily "cleans up" chemically in the laboratory.

    (b) only has positive effects.

    (c) activates receptors required for therapeutic effect and others which cause side effects.

    (d) only has adverse effects.

(5) The "A" (as in SDA or NaSSA), written in a drug's mechanism of action (MOA), stands for _____.

    (a) agonist

    (b) antagonist

    (c) inverse agonist

    (d) partial agonist

    (e) all the above

ANSWERS

(1) The correct answer is both "b" and "d." "B" is Latin for mechanism of action. Bonus credit if both were selected.

(2) TRUE. However, give yourself credit if you marked FALSE. All drugs have four names: chemical, generic, brand, and street. But throw in funny, slang nicknames that prescribers use (and some very astute patients), then they have five.

(3) TRUE. That is a fact. No exception. They kick in with the very first dose.

(4) The correct answer is "c" – works where you want it to work and works where you do not want it to work. Working where one does not want it to work yields side effects.

(5) The correct answer is "e," all the above. The letter "A" written in a drug's mechanism of action is a wildcard. It may stand for any of these; usually, however, the "A" is most often written to mean either "agonist" or "antagonist."

Module 5

Winner Takes All

Pharmacokinetics

☺ ☺ ☺ ☺

With an understanding of a drug's mechanism of action (MOA), it is time to get down to other basics. It all starts with where and how the drug goes in, where the drug goes from there, what the drug does when it gets there, and how it exits the body — pharmacokinetics and pharmacodynamics. It is a prizefight between kinetics and dynamics with a big prize, and the winner takes all — first, pharmacokinetics.

Think of pharmacokinetics (PK) and pharmacodynamics (PD) as two fighters in the ring, competing against each other for the prize. Pharmacokinetics is a wrestler named "Your Body," and pharmacodynamics is the competitor named "The Drug." They fight. Sometimes the body wins. Spoiler alert: Most of the time the drug wins (because they have the backing of the big drug companies who go out of their way to make a drug that tricks "Your Body" into throwing in the towel early). Or put another way, "Your Body's" job is to deflect any foreign punches and protect itself from any real or perceived harm (the body does not know the difference). The drug's job is to deliver as many sucker punches as it can, sneak around the body's lethal guards, and enter the body's main highway – the bloodstream. The kidneys get the last laugh. The kidneys always get the last laugh.

Why is it "winner takes all?" If pharmacokinetics ("Your Body") wins, the drug never gets in and is eliminated straight away. If pharmacodynamics ("The Drug") wins, the drug gets into your systemic circulation and rapidly takes over. Admittedly one wants the drug to find its way to the source of one's illness and fix it, but thinking of pharmacokinetics and pharmacodynamics as antagonistic to each other makes it easier to understand the function of both.

And the bell rings. The battle between body and drug begins. The first fighter, "Your Body," enters the ring.

Pharmacokinetics is <u>what the **body** does to the drug</u>. Take the word, dissect it, and "pharmaco" (from Greek, "pharmakon") means "drug" or "poison," and "kinetics" (from Greek, "kinetikos") means to "to move." Put these two together, and it means a "drugs movement through the body." At each point in the drug's movement, the body tries to deflect "any drug punches" that could ultimately harm the body.

Pharmacokinetic's goal is to get the alien drug, (also known as a xenobiotic), out as quickly as it can. The body's default assumption is that "something ingested" is a poisonous stranger, the xenobiotic, until proven innocent.

There are four major divisions within pharmacokinetics and are associated with the mnemonic "A-D-M-E" (pronounced "ADD-ME"). They are "Absorption, Distribution, Metabolism, and Elimination." There are further subdivisions and associated topics like bioavailability, biotransformation, and the microsomal mixed-function oxidase system (also known as the "CYP" system). Here is how "A-D-M-E" and upcoming "A-D-M-E-T" works.

The drug (food or anything put into the mouth and then swallowed) goes in and is absorbed into the system. The bloodstream distributes the drug, and liver metabolizes or breaks it down into something safe or toxic. Finally, the kidneys, the most unforgiving organ in the body, excrete. No matter whether it is friend or foe, the kidneys kick it out.

Time to add another letter to "A-D-M-E," one that is rarely added to this mnemonic. Make it "A-D-M-E-T" (pronounced "Add-MET"). This "T" stands for "time." Sometimes "T" can mean toxicity; however, in most texts, "T" stands for "time." The amount of time (also known as half-life and abbreviated "t½") is how long the drug stays in the body, how long it takes to reach a therapeutic level or "steady state," and how fast it is broken down and cleared from the body. A healthy liver and kidneys manage this efficiently; a diseased system will have trouble breaking down and clearing the medication. Giving "time" a prominent place in this mnemonic squarely demonstrates the power of pharmacokinetics. Without the time component, pharmacodynamics, the other fighter, would run amuck.

Pharmacodynamics (from Greek, "pharmakon" and "dynamis" means "power") is what the drug does to the body. Pharmacodynamics is the drug's power over the body. Here is an easy way to remember that. Note that pharmaco<u>d</u>ynamics has a "D" at the beginning of the fourth syllable of the word. The word "drug" also begins with the letter "D." One can easily remember that Pharmaco**D**ynamics is what the **d**rug does to the body. This point will be repeated again and again; there are a multitude of actions involved in pharmacodynamics, and they will be discussed in later modules.

This fighter's (the drug) goal is to "trick" the body into allowing as much of the drug as possible into the system. The drug companies always create a worthy adversary, and it usually gets in (especially if the liver is not working at full strength). Once the drug has entered the system, numerous terms are associated with pharmacodynamic actions – descriptors like agonist, antagonist, partial agonist, indirect agonist, inverse agonist, potency, efficacy, dose-response curve, ED_{50}, TD_{50} (also known as LD_{50}), therapeutic index, and more. If these words seem like a foreign language, not to worry, explanations follow.

What is this module's takeaway? When you think "pharmacokinetics," think "body." When you think "pharmacodynamics," think "drug." Pharmacokinetics is the body's way of dealing with something initially perceived to be foreign. Pharmacodynamics is the drug's way of manipulating the body's system.

QUIZ TIME

(1) True or False? All drugs have four names: chemical, brand, generic, and street.

(2) True or False? Pharmacokinetics is the pathway of the drug through the body.

(3) True or False? Pharmacodynamics is the "power" of the drug after it is allowed into the system.

(4) True or False? All risks associated with pharmacogenomics or pharmacogenetics have been identified.

(5) True or False? The concept of pharmacokinetic time is often left out of the "A-D-M-E" mnemonic.

ANSWERS

(1) TRUE. Street names are the most difficult, because they are constantly changing.

(2) TRUE. From the mouth to anus, and all points in between, pharmacokinetics is the body's way of moving and inactivating a medication. In the next module, the routes by which drugs are absorbed will be explained.

(3) TRUE. The power of the drug can vary, but what the drug does to the body can only be viewed as its power over the body.

(4) FALSE. Presently, all anyone can do is conjecture. It is suspected that new pharmacogenomic risks will emerge as the field, knowledge base, and society continue to progress.

(5) TRUE. The first question a patient will always ask: How long before the drug works? The importance of medications and timing cannot be understated – and it often is.

Module 6

The Right Way

Routes of Administration

☺ ☺ ☺ ☺

If the first "A" of pharmacokinetics is "Absorption," then the precursor "A" that must happen before any absorption is another "A," administration. Absorption is dependent on the method of administration. Not to be overly technical, but absorption is not only dependent on the method of administration, but the drug's solubility (ability to dissolve) and ionization (charge), along with an individual's age, gender, and body size. More on this later; back to administration. How are drugs administered? One may take it, stick it, suck it, push it, sniff it, punch it, prick it, slide it, rub it, stab it, jab it, attach it, puncture it, and/or chew it.

As a rule, if a body orifice is available to "accept" a medication, it is used. If an orifice is not available, an artificial opening can be made. There are two main routes of administration or pathways that determine how much of the drug reaches its target and how quickly it gets to that site. They are enteral and parenteral, and each provides distinct advantages and disadvantages.

An enteral route of administration is referred to as "enteral," not because it "enters through the mouth," but because it will enter the mouth and move through to the "enteric system." Most people do not know they have an enteric system.

Occasionally they may notice that a medication is "enterically coated," but they rarely think twice about what that means. To understand the "enteral pathway," first look at the word "enteric." Aristotle was the first to use it, and it comes from the Latinized form of the Greek word "enterikos," meaning "intestinal." After 1822, it was used in everyday medical practice to mean "intestines." Put it all together, and it means "in the intestines." If you take an "enterically coated medication," it is designed to be protected from stomach acids and absorbed more in the intestines that in one's stomach.

The mouth is the most straightforward enteric pathway – and remember, it is not called "enteric" because it "enters the mouth." Enteric means it goes through the gastrointestinal system – notably the liver. With the "enteric pathway," think liver. The parenteral pathway bypasses the liver.

Take it. The pill goes in your mouth. What happens next? Few think further about this – including medical and patient types. It goes from your mouth to throat; throat to your stomach; stomach to intestines; intestines via the hepatic portal vein to the liver; liver to the blood stream; blood to receptor sites throughout the entire body. When the drug does its job, it returns to the venous circulation on to the kidneys, then out. This is certainly an oversimplification. Anywhere along this pathway, a drug can be absorbed into the system; but overall, this is the route of a drug when taken by mouth, enterally. From mouth to liver takes about twenty minutes. As a result, a limiting factor of enteral medications is the time it takes to be absorbed.

One important point to remember: If a drug is enteral and goes through the liver, this is called "first pass." This means that an enteral route of administration allows the liver, the gatekeeper, a lethal guard, a "first pass" or first chance to get rid of the drug. That is the liver's function. It is the gatekeeper, designed to keep harmful things out. This "first pass," is the body's first attempt to protect the entire system.

Stick it. The other pathway considered enteral, although it can have elements of both enteral and parental pathways, is rectal administration (suppository). The rectum is a muscular tube connected to the end of a person's colon. In the rectal area, portions of the drug absorb through the hemorrhoidal vein, avoid some first-pass administration, and make its way into systemic circulation. This route is often irregular and incomplete. If it is pushed out of the rectal mucosa into the large intestine (technically referred to as deeper placement), then more of the drug will be absorbed into the large intestine, and more of the drug will make its way to the liver. That explains why medical professionals are taught how to "stick or slide it." They are advised: "Never push a suppository further than your finger's second knuckle" (beyond that equals deeper placement into the large colon). If inserted too far, then much more of the drug is broken down. It will now go into the hepatic portal vein system; the liver will break it down, and the drug will be unable to do its job.

The second route of administration is "parenteral." Enteric means "intestines," and the prefix, "par," or "para" means "along-side, near, or around." Put it together, and it means "around the enteric system" or "around the liver." This difference between "enteral" (through the liver) vs "parenteral" (around the liver) is a fundamental distinction to remember. If it goes around the liver, the gatekeeper, toxicities and overdosing can happen more quickly. The liver can impact a drug's entrance into the systemic circulation – slowing it, reducing it, or even eradicating it. If the drug bypasses the liver, parenterally, the drug will go straight to the bloodstream while "thumbing its nose" at the liver. So, enteral is slower, safer, easier, and economical. Parenteral is faster with limited body intervention and protection, or put another way, dangerous. It is hard to reduce a dose once it has been administered parenterally into the system.

Here are other, but not all common parental routes of administration that <u>bypass</u> the liver.

Intravenous (IV). Push it. Entrance directly into the bloodstream is rapid and provides the most accurate blood concentration; but administration is hard to reverse, requires sterile needles, and is considered a medical technique.

Intramuscular (IM). Jab it. Utilizes a hypodermic needle, slow with fairly even absorption; but localized irritation is possible and medical administration required.

Subcutaneous (SC). Prick it. The drug goes under the skin using a short needle injected between the skin and muscle. Slow and prolonged absorption, but variable absorption, depending on blood flow at placement site.

Topical. Rub it. Localized action, easy to administer; but may be absorbed into greater circulation, so it may not stay local.

Transdermal. Attach it. Active ingredients are absorbed and delivered via skin patch or ointment for systemic circulation. The good news is that absorption can be controlled and prolonged, but local irritation is possible.

Intranasal. Sniff it. Easy, rapid, local effects, bypasses the Blood Brain Barrier (BBB), but not every drug can be turned into a spray, plus potential irritation of nasal mucosa.

Epidural. Puncture it. Gently. A local anesthetic is injected in the lower back into the space outside the dura mater of spinal cord, enters the brain bypassing the blood brain barrier (BBB). Immediate effects, notably in the central nervous system (brain and spinal cord), but not reversible, possible nerve damage, and requires trained personnel to administer.

Bottom line: Whether one takes it, sticks it, sucks it, sniffs it, pushes it, smells it, punches it, pricks it, slides it, rubs it, stabs it, jabs it, attaches it, or punctures it, the drug goes in. Whether it can remain, that is another question altogether.

QUIZ TIME

(1) If a drug is administered parenterally, it bypasses the _____.

      (a) mouth

      (b) nose

      (c) kidneys

      (d) liver

(2) The two enteral routes of administration are _____.

      (a) IV and IM

      (b) transdermal and oral

      (c) oral and rectal

      (d) rectally and epidurally

(3) The most significant factor to consider when deciding to administer a drug enterally versus parenterally is _____.

      (a) ease of administration

      (b) cost of medications

      (c) medical supervision

      (d) safety

(4) Before absorption, there must be _____.

      (a) distribution

      (b) metabolism

      (c) administration

      (d) insurance payments

(5) Pharmacokinetics and pharmacodynamics explain what the _____ does to the drug and what the _____ does to the body, respectively.

    (a) body; drug

    (b) drug; body

    (c) drug; brain

    (d) body; brain

ANSWERS

(1) The answer is "d," liver. Parenteral administration bypasses the liver. Enteral, first pass, the drug goes through the liver.

(2) The answer is "c," oral and rectal.

(3) The answer is "d." Perhaps some limited debate here. All choices are worthy of consideration, but maintaining safety is the key consideration.

(4) The answer is "c." No administration, the drug never enters the system, then there is no absorption, distribution, metabolism, or elimination. No administration; no "A-D-M-E."

(5) The answer is "a." Pharmacokinetics is what the body does to the drug. Pharmacodynamics is what the drug does to the body. Remember, pharmacodynamics has a "D" in it, and that stands for "Drug."

"Abandon All Hope, Ye Who Enter Here"
—Said the Liver to the Drug
Originally from Dante Alighieri's *Inferno* (1314)

Absorption and Distribution

☺ ☺ ☺ ☺

Absorption and distribution of a drug through the blood system into the body is typically taken for granted by the client or patient, and it is not as simple as it sounds. The route of administration can alter the rate of absorption and distribution. Blood levels of the same drug dose administered using different routes (enterally and parenterally) can vary significantly. Other factors that can influence drug absorption include: Area of absorbing surface, number of cell layers between the site of administration and the blood stream, degree of drug destroyed by liver (biotransformation), digestion, and the degree of binding to food or inert complexes within the blood. Additionally, absorption can be affected by age, gender, body size, personal genetics, pain, stress, hunger, food (hot/solid/fatty foods delay absorption), fasting, exercise (decreases blood circulation to stomach and reduces absorption), and other body and age differences. Finally, the chemical makeup of the drug (e.g., solubility and ionization) certainly affects absorption. Admittedly, this is complicated; nevertheless, both absorption and distribution are best understood from the framework of a crucial word: Bioavailability, the percentage of drug dose that makes its way to the blood stream.

Think of absorption as the process of delivering a drug to the blood stream, and distribution as the highway ("blood stream") that takes the medication to the remote parts of the body. To put it another way: Drug absorption is a medication's movement from the site of administration into the bloodstream. Distribution delivers a drug via the bloodstream to the tissues of the body. One final metaphor: Hopping in your automobile (administration) and making your way describes absorption; driving your vehicle down the highway represents distribution; and arriving at grandma's house is the site of drug operation. Getting there safely and intact is bioavailability. The liver is the traffic cop.

The pill enters (enteral administration) and moves from mouth to throat; throat to stomach; stomach to small intestines; intestines via the hepatic portal vein to the liver and that is where the bulk of bioavailability transpires. The liver cop or guard pulls the drug over and starts breaking things down (metabolism).

Three words to remember: Liver, metabolism, and bioavailability. The liver is a body's best friend, and it is not about to let anything into the system that could cause harm. The first job for the liver is metabolism (from Greek, "metabole," meaning "change"). The liver's job is to biotransform or simply break down the drug. The best synonym for biotransformation is inactivation — although sometimes the liver attempts to inactivate and instead accidentally activates. Mostly, the liver gets it right, but sometimes the liver gets it not quite so right. So, when the liver comes in close contact with a drug, it can (1) do nothing and grant admission, (2) biotransform it into an active or inactive metabolite, or (3) break it down into a toxic metabolite.

Whatever the result, the drug or its metabolite enters the bloodstream on its way to the job site. When it gets past the liver, pharmacodynamics is in control.

The pill enters via the mouth (approximately 80% of all drugs are administered this way), down the throat, into the stomach, small intestines, and liver via the hepatic portal vein. Once the liver has its chance at first pass and "releases" the drug or its biotransformed product into the bloodstream, the drug moves to the job site, does its job, then enters venous circulation, and now the kidneys do their part.

The kidneys are unforgiving, and it is vital that they stay that way. They have one job — clear or eliminate. When one thinks kidneys, the word that should come to mind is clearance; and to reiterate, when one thinks liver, the two words that should come to mind are biotransform and metabolize.

If bioavailability is associated with absorption and distribution, then biotransformation is the word "du jour" for metabolism and clearance. Also, time to add in the "T" in "A-D-M-E." The "T" stands for time or "half-life."

There are three words to remember concerning drug clearance: half-life (abbreviated $t\frac{1}{2}$, with "t" standing for "time"), first-order kinetics, and zero-order kinetics.

First off, half-life or $t\frac{1}{2}$. The standard definition found in any pharmacology textbook states that the half-life of a drug is the time for the drug in the body's plasma concentration to reduce half its original value. In other words, half-life is an estimate of how long it takes for half of the drug to be removed, cleared, or eliminated from the systemic circulation.

Half-life determines dosing and is calculated by the pharmaceutical manufacturer. If a drug must be taken every four hours, then its half-life is four hours. If a drug's required dosage is twelve hours, then that is the half-life. If a drug must be taken every twenty-four hours, then twenty-four hours is the half-life. One important point to remember, the longer the half-life, the easier it is to remember to take the drug and increase the likelihood of patient compliance. In other words, patients are more likely to comply with prescriptive orders to take a drug once a day than they are to take a medication at varying times during one twenty-four-hour period.

As these drug molecules find their way through the system, the kidneys clear them. If this did not happen, then these drugs would re-enter systemic circulation and indirectly raise the dose concentration when the next drug dose is taken. Most drugs are cleared via first-order kinetics. The keywords to remember are proportion and fraction (50%). Referred to as exponential elimination, a constant fraction (50%) of free drug in the blood is removed during each half-life time interval. To explain each half-life, half of the drug is cleared, then half of what is left, then half of that half, then half of the remainder. From 100%, (1) 50% is cleared in the first half-life, then (2) half of that 50% or 25% is removed, (3) then 12.5% is cleared during the next half-life, (4) then 6.25%, (5) 3.125% and so on until the drug is completely cleared from the system. As a rule, it takes 5 to 5.5 half-lives for a drug to be removed from the body. After six half-lives, it is virtually eliminated.

Here is an example that puts all this together. Ambien®/zolpidem (brand/generic, notice no caps on generic) is a popular sleep aid. It has a half-life of about two hours and is a right choice for someone who has trouble getting to sleep (sleep latency) but has no difficulty staying asleep. Multiply two hours (the half-life of Ambien®/zolpidem) times 5.5 half-lives, and it takes approximately eleven hours for the majority of this drug to be eliminated or cleared or removed from the body.

It is therapeutically essential that a constant amount of the drug be maintained in the system. This is referred to as "steady state plasma level" or when the absorption/distribution phase is equal to the metabolism/excretion phase. If an overly technical definition is preferred, "steady state" is when the substrate is being formed at the same rate it is being broken down. Scientific language means the amount of the drug going in and circulating the body is equal to the amount of the broken down drug. How long does it take to reach steady state? Five doses. After the fifth dose, and if one continues to take the medication as prescribed, it will maintain a therapeutic, beneficial level in the system. Now, put the two together. It takes 5.5 half-lives to reach steady state, assuming first-order kinetics. Discontinue the drug, and it takes 5.5 half-lives to clear from the system.

Imagine the following scenario. A bacterial infection is afoot, and bacteria have established a quorum or foothold in the system. They are alive and well and building a sickening population in unwanted places. The prescriber says, "Take these pills, one every eight hours, and do not stop until you have taken all of them." Why? If you take one pill, you have achieved some small success , but left a vast, sufficient number of bacteria to do their job and maintain the sickness. In this story, the half-life is eight hours. After five doses, or roughly 40 hours, a steady, therapeutic level is achieved. Now, bacteria do not have a chance. Enough medication is present to kill and to stop them in their little bacteria tracks. After a patient has taken the entire bottle of pills, it will take roughly five half-lives for the drug to clear the system. In this case, assuming first-order kinetics, after approximately forty hours, the drug has been cleared from the system. The clinical effect is over. It takes 5.5 half-lives to reach steady state; then assuming all organ systems are working properly, 5.5 half-lives of no further drug administration will clear the drug from the system.

Elimination varies from person-to-person based on their age, weight, other drug interactions, medical conditions, liver function, and kidney function. Therefore, $t\frac{1}{2}$ (half-life) estimates how long it takes to remove the drug from the body. It is not always about the liver and kidneys. Drugs can also be eliminated via saliva, breath, breast milk, and sweat. If the drug is administered parenterally, the liver's first pass is sidestepped, but kidneys always clear.

If the keyword to remember with first-order kinetics is proportion and time, then the keyword to remember with zero-order kinetics is amount – a fixed amount. Most drugs are cleared from the system by first-order kinetics; but at least one very well-known drug is cleared by zero-order kinetics.

Zero-order kinetics means that drug molecules are cleared at a predictable, constant rate, per unit of time, independent of the concentration of drug in the system. The best example of zero-order kinetics is ethyl alcohol. Imagine enjoying an "adult beverage." Alcohol is metabolized at approximately 10 to 15 ml/hour, or 1.0 ounce of 100-proof alcohol per hour regardless of concentration. To put this in perspective, beer is approximately 9-proof; wine is 27-proof; whiskey is 86-proof. For the sake of example, imagine you are having a drink closer to 100-proof. It will take one hour per 1.0 ounce of alcohol to clear your system. If you have three drinks, think three hours for it to get out. Four drinks, four hours (Meyer & Quenzer, 2019).

What happens when alcohol levels are elevated and routes of metabolism or elimination are saturated (that means, there is more drug or alcohol in the system than possible sites or enzymes to break it down). Alcohol roams around in the bloodstream and waits for its turn to be broken down. Depending on how much one imbibes, blood alcohol concentration (BAC) increases. When two or more drinks are guzzled in a short period (think less than an hour), and metabolism occurs at a fixed rate, stop drinking. Any additional consumption will raise blood levels dramatically, and, of course, produce intoxication. Alcohol breakdown starts with zero-order biotransformation when there are high alcohol levels in the system then shifts to first-order kinetics as blood levels reduce. Other common medications broken down by zero-order kinetics include Dilantin®/phenytoin, Coumadin®/warfarin, heparin, aspirin, and tolbutamide.

One distinct advantage to zero-order kinetics: It is defined from the start exactly how long the medication will stay in the system.

HALF-LIFE & STEADY STATE

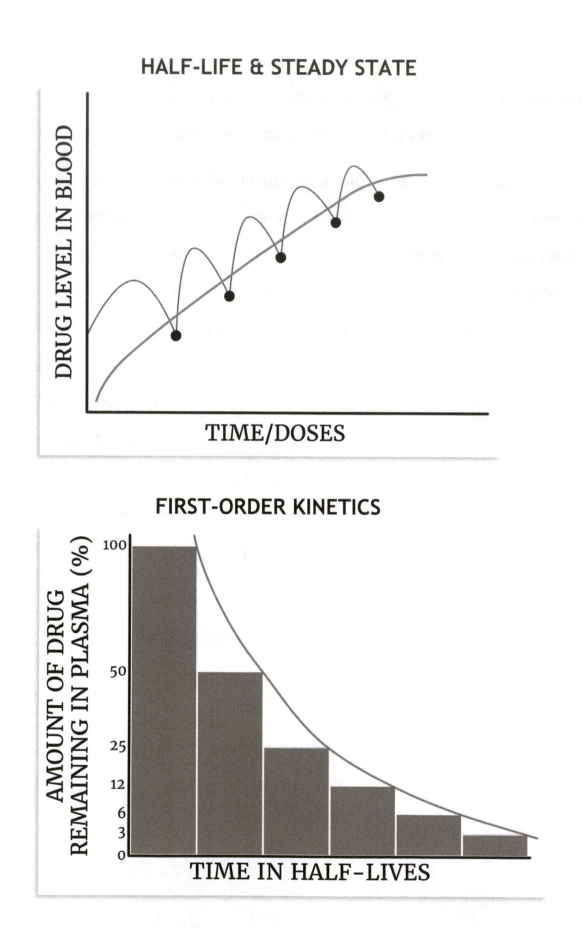

FIRST-ORDER KINETICS

QUIZ TIME

(1) Before absorption, there must be _____.

    (a) distribution

    (b) metabolism

    (c) administration

    (d) elimination

(2) _____ percent of drugs are administered parenterally.

    (a) 10%

    (b) 15%

    (c) 20%

    (d) 25%

(3) Another word for "biotransformation" is _____.

    (a) inactivation

    (b) activation

    (c) bioavailability

    (d) clearance

(4) Assuming first-order kinetics, when a drug is taken as prescribed, after _____ doses, a therapeutic, steady state is achieved.

    (a) 4.5

    (b) 5.5

    (c) 6

    (d) 6.5

(5) If 100% of a drug is found in your urine, how bioavailable was that drug?

    (a) 0%

    (b) 25%

    (c) 75%

    (d) 100%

ANSWERS

(1) The answer is "c." No administration, the drug never enters the system. No absorption, distribution, metabolism, or elimination. No administration; no "A-D-M-E."

(2) The answer is "c," 20%. If 80% of all medications are administered enterally (by mouth or rectally), this simple subtraction yields 20% parenterally.

(3) The answer is "a," inactivation, to render the drug harmless.

(4) The answer is "b," 5.5 is the number of half-lives required to reach steady state. Remember, once the medication is no longer needed, after 5.5 half-lives have passed, the drug is out of the system.

(5) TRICK QUESTION. The answer is "d," 100%. If the liver "allows" 100% of a drug to enter the bloodstream, then 100% of that drug makes it way to the kidneys, which filter 100% of the drug molecules out. In that case, the drug is 100% bioavailable. If 100% of a drug is found in a patient's fecal matter, then the drug would be 0% bioavailable.

The Super Family

CYP450 Enzymes

😊😊

Meet "The Super Family" of fifty or more enzymes — The CYPs. They are also known as the CYP450 Oxidase Enzyme family. And if that is not enough to remember, they also have a more formal name: The microsomal mixed function oxidase system. Referring to this super family as the "C-Y-P 450 System" (saying each letter) is correct. Calling them the CYP450 (pronounced "sip" four-fifty) System is equally correct. However, those in the know refer to this super family as the "Cytochrome P450 (pronounced P-four-fifty) System."

How did they get their "strange" "peculiar-sounding" name? Break it down: Cyto-chrome-P450. CYP450 enzymes are named because they are bound to cell membranes (that is the "cyto" part); contain a heme pigment (that is the "chrome" and "P" part); and absorb light at a wavelength of 450 nm (nanometers) when exposed to carbon monoxide. (Petrović, et al., 2020).

How did the individual, specific enzymes or members of the CYP family get their name? Each family member within the super family follows the following naming formula: A number, followed by a capital letter, followed by a number. So, the most common CYP enzymes are 3A4, 2D6, 2C9, and 2C19. In 3A4, the "3" is the family number; the capital letter that follows, in this example, "A," is the "subfamily letter," and the number at the end is the number for the isoform or individual enzyme. They are classified into families and subfamilies by their amino acid sequence and by the gene encoding them.

CYP450 enzymes were not discovered until 1954, when Martin Klingenberg first discovered them during his research on steroid hormone metabolism when he extracted a novel protein from hepatocytes (Klingenberg, 1958). Years passed until other researchers (Cooper, Levin, Narasimhulu, Rosenthal, and Estabrook) discovered their power as a key enzyme in drugs and steroid hydroxylation (Estabrook, 2003). However, another twenty-plus years passed before their power in drug interactions was discovered; and now, with modern-day pharmacogenomics (PGx), they are finding a prominent place in prescribing (McDonnell & Dang, 2013).

CYP450 enzymes, just like other enzymes, are molecules that make chemical reactions more efficient. While the following may sound complicated and esoteric, it is important to have a basic understanding of the complexity of this super family. To that point, CYP enzymes are versatile "hemeproteins or hemoproteins." Another example of "hemeproteins" is hemoglobin which carries oxygen in the blood and is present in the mitochondria and the endoplasmic reticulum (found in eukaryotic cells; and humans are categorically eukaryotic organisms).

Hemeproteins are essential for every organism and serve several vital functions. They are capable of "hosting" iron molecules. They carry oxygen and transport or store electrons. One cannot live without them.

CYP enzymes are found in all kingdoms of life: animal, plant, fungi, protists (pronounced, "pro-tusts"), bacteria, and archaea (pronounced "are-kee-uh"). Also, viruses, not living, have CYP450 enzymes. By far most of these enzymes make their home in the liver. They can be found, although not in abundance, in lungs, intestines, kidneys, nasal passages, and plasma (MdConnell & Dang, 2013).

If the liver is the gatekeeper of your system, then the CYP enzymes are the guards. CYP450 enzymes represent a protective biochemical process whose job is to break down and oxidize "xenobiotics" – foreign substances in the body. Xenobiotics (from Greek, "xenos," meaning "foreigner or stranger"), can be grouped as carcinogens, drugs, environmental pollutants, food additives, hydrocarbons, pesticides, and anything that is foreign and deserves to be evaluated for its potential to cause harm.

While the four most common CYP450 enzymes are "3A4," "2D6," "2C9" and "2C19," it is "3A4" and "2D6" that oxidize, metabolize or break down, one-half of all oral medications. The "3A4" enzymes break down most drugs in the entire pharmacopeia, but "2D6" enzymes break down most of the psychotropic medications and are the most extensively studied. For the sake of trivia, "1A2" breaks down caffeine; "2E1" helps metabolize alcohol; "2C9" breaks down Coumadin®/warfarin. At their simplest level, they keep you alive and well.

QUIZ TIME

(1) Which of the following is a CYP enzyme?

   (a) 3A4

   (b) BQ7

   (c) #A4

   (d) AGCT

(2) Where in the body will one find CYP enzymes?

   (a) Lungs

   (b) Kidneys

   (c) Intestines

   (d) Plasma

   (e) All the above

(3) If zero percent of a drug is found in your urine, how bioavailable was that drug?

   (a) 0%

   (b) 30%

   (c) 60%

   (d) 90%

(4) Which CYP450 enzyme breaks down the most psychotropic medications?

   (a) 2C19

   (b) 2E1

   (c) 2C9

   (d) 2D6

(5) What is the purpose of CYP enzymes?

   (a) Make chemical reactions more efficient

   (b) Break down xenobiotics

   (c) Turn drugs into soluble molecules

   (d) Function as a protective biochemical process

   (e) All the above

ANSWERS

(1) The correct answer is "a." 3A4 follows the formula: Number, capital letter, and number.

(2) The correct answer is "e," all the above, and robustly in the liver.

(3) The correct answer is "a." This drug is zero percent bioavailable. It never made it past the enzymes in the liver. CYP strikes again.

(4) The correct answer is "d." The 2D6 enzyme breaks down the most psychotropics; the 3A4 enzyme breaks down most drugs in the pharmacopeia.

(5) The correct answer is "e," all the above. CYPs protect from harm – or at least they do their best. They biotransform. Their action explains why drug effects wear off. As guards, they do not allow perceived harm to enter the system. Spoiler alert: They do, indeed, turn drugs into soluble molecules. Now you know what the next module is all about.

Module 9

Breaking Up Is Hard to Do

CYP450 Enzymes Redux

☺

First off, the bottom line, up front. CYP450 Enzymes transform lipophilic, "fat loving" things into hydrophilic, "water-loving" things. That is the final product of CYP450's hard work. These two words – lipophilic and hydrophilic – represent the key underpinnings of pharmacology and psychopharmacology. If a drug is lipophilic, it can be absorbed into the system. If a drug is hydrophilic, it cannot be absorbed into the system. What is the easiest way to get a drug out of the system? Take something that is fat-loving (lipophilic or hydrophobic) and "magically," through biochemistry, make it soluble, water-loving (hydrophilic or lipophobic). How does this happen? Give it a charge, turn it into a hydrophilic, polar molecule also known as an ionized molecule, and the kidneys will excrete it right out. Get these words straight, along with their synonyms, and that is all the biochemistry you need to know – for right now.

LIPOPHILIC VS HYDROPHILIC

LIPOPHILIC	HYDROPHILIC
Facilitates Absorption	Facilitates Excretion
Beginning Product	End Product
Insoluble	Soluble
Fat Loving	Water Loving
Hydrophobic	Lipophobic
Nonionized	Ionized
Nonpolar	Polar
Active	Non-Active
Associated or Integrated	Dissociated
Unprotonated	Protonated
Uncharged	Charged
Absorbed	Excreted
Easy to get through membrane	Hard to get through membrane
"Keep It In"	"Get It Out"

There are two major ways that liver microsomal enzymes biotransform (inactivate). Remember, the liver biotransforms efficiently, and chemical changes are also enzymatically catalyzed in many tissue organs including stomach, intestines, blood, lungs, nasal passages, kidneys, and brain. But overall, the liver wins the biotransformation blue ribbon.

These two major types of biotransformation are called (1) Type I or Phase I and (2) Type II or Phase II. The reason Type I is also called Phase I is because these chemical reactions can occur before there is a second step or Phase II. In other words, a drug's breakdown likely starts with Phase I and may or may not end up requiring Phase II. Put another way: Some drugs, like phenobarbital, only need Phase I; others, like morphine, only need Phase II. Then there are drugs like aspirin that require both Phase I and Phase II to be metabolized, or an "oddball" like Valium®/diazepam which goes through Phase I twice, and then goes through Phase II. Rarely will a drug start with Phase II and then go to Phase I, but isoniazid (isonicotinyl hydrazide; INH) for tuberculosis does just that (Wang, et al., 2016).

Phase I (also called "functionalization") is the classic chemistry "REDOX" (an oxidation-reduction reaction that involves a transfer of electrons). The result is a broken-down drug product (metabolite) which once was lipophilic but now is hydrophilic. Put another way, now the drug has been biotransformed, inactivated, turned into something harmless (not always) and can be excreted by the kidneys and shown the door. This is a simplification, but that is the goal.

Type I, also called Phase I, and referred to as "functionalization," utilizes the CYP enzymes. CYP enzymes are responsible for oxidizing most psychotropic medications including antidepressants, amphetamines, morphine, and antipsychotics.

Type II, also called Phase II, and referred to as "conjugation," works differently and does not involve CYP enzymes. It takes a nonpolar, lipophilic molecule and biotransforms (inactivates) it into a polar, charged molecule, one that is water loving and easily moves through the kidneys and out the door.

Phase II is a synthetic (Phase I is non-synthetic) reaction that combines (conjugates) a drug with some small molecule like glucuronide (particularly important for psychotropic medications) or sulfate. The result: Once lipid and loved, now water soluble and banished. The result is an ionized and likely biologically inactive product, ready to be shown the door.

Phase I and Phase II work differently but have the same goal: Biotransform, inactivate, make a drug more water soluble, turn it into something that is ionized or charged, render it biologically inactive, and wash it out of the system.

If the pill is administered enterally, it goes throat to stomach, stomach to intestines, intestine via the hepatic portal vein to the liver. That is where most of this process of breaking up and throwing out commences. As a result, the liver can either (1) do nothing to a drug, just let it in, (2) turn it into an active metabolite, (3) biotransform it into an inactive metabolite, (4) turn part of it into a toxic metabolite (acetaminophen is an example here; 5%-10% of the drug is transformed into a toxic metabolite), and/or (5) break the whole thing down and start the exit process (not likely, what drug manufacturer would allow a drug to come to market that was metabolized completely and therefore served no purpose) (Flint, et al., 2017; McGill & Jaeschke, 2013).

After the liver finishes with the drug, the byproduct (active metabolite, inactive metabolite, toxic metabolite, or no drug breakdown) is "free" to roam through your system, do what it is supposed to do (e.g., bind to a receptor), return to the venous circulation where it heads to the kidneys, and then the kidneys clear the drug from the system. Once upon a time, the drug was lipophilic and absorbable. Now it is hydrophilic, and the kidneys are able to clear the drug from the system.

One final point, when it comes to drug metabolism and clearance, the two most important organs in the body are the liver and kidneys, respectively. Healthy liver and kidneys are prized possessions. However, as the body ages, the liver and kidneys lose about 1% (more in the range of .70%) of their ability to break down and to clear each year after the age of thirty. This is not the result of disease which could exacerbate the problem, this 1% loss is just natural aging (Denic, et al., 2016; Li, et al., 2012). Happy Birthday.

The moral of the story? Take care of your liver and kidneys, and they will take care of you.

QUIZ TIME

(1) Which of the following is a synonym for biotransformation?

      (a) bioavailability

      (b) biologic

      (c) bioactive

      (d) inactivation

(2) True or False? It is common for drugs to go through Phase II and then return to Phase I.

(3) The liver _____, and the kidneys _____.

      (a) metabolizes/biotransform

      (b) metabolizes/clear

      (c) clears/metabolize

      (d) clears/bioavail

(4) Which CYP450 enzyme breaks down the most psychotropic medications?

      (a) 3A4

      (b) 2C9

      (c) 1A2

      (d) 2D6

(5) CYP450 enzymes are associated with _____.

      (a) Phase I

      (b) Phase II

      (c) Conjugation

      (d) Phase III

ANSWERS

(1) The answer is "d," inactivation.

(2) FALSE. It happens, but it is uncommon for a medication to go through Phase II then return to Phase I.

(3) The correct answer is "b." Worth repeating: The liver metabolizes, and the kidneys clear.

(4) The correct answer is "d." 2D6 breaks down the most psychotropics, and 3A4 breaks down the most drugs in the entire pharmacopeia.

(5) The correct answer is "a." When you think CYP450, think Phase I, functionalization.

Module 10

Special Handling

Prodrugs

☺ ☺ ☺ ☺ ☺

There is another important layer concerning Phase I (functionalization/CYP enzymes) and Phase II (conjugation) upcoming in Module 11, CYP450 inducers and inhibitors; but for the moment, here is something easy to "metabolize" and rather curious.

There is a class of medications that requires special handling by the liver. To that point, they are administered via an enteral pathway (by mouth) and head straight to the liver for superior handling. They are "prodrugs."

A prodrug is biologically inactive when taken, then the liver works its magic (either through CYP or conjugation) and metabolizes it into an active compound. It is inactive when taken and activated by the liver. It must be an enteral medication that undergoes a chemical reaction in the liver (first pass) thereby becoming an active pharmacological agent. A parenteral drug can never be a prodrug because parenteral pathways (IV, IM, sublingual, subcutaneous, transdermal, and others) bypass the liver. A prodrug requires first pass to turn an inactive drug into an active drug moiety (pronounced: "moy-uh-tee"). This active molecule or ion (not including inactive portions) produces a physiological or pharmacological response (Wu, et al., 2009).

A prodrug is easy to understand. When taken, it is inactive, does nothing. The liver, via first pass, utilizes either Phase I or Phase II metabolism to turn this inactive substance into something that has a medicinal effect on the body. In other words, a prodrug is a precursor – something that comes before, required, essential to make something new. For instance, lumber is a precursor in the building of a house. Proper nutrition is a precursor in maintaining good health. Protein is essential in the creation of good mental health. Parents are the precursors of a child.

Prodrugs are common, and the prototypic prodrug (representing an original type to which other things are compared) is codeine. See the following chart.

PRODRUGS	
INACTIVE PRODRUG → (BIOTRANSFORMED INTO) → ACTIVE METABOLITE	
PRODRUG	ACTIVE METABOLITE
codeine	morphine
aspirin	salicylic acid
Risperdal®/risperidone	Invega®/paliperidone
Effexor®/venlafaxine	Pristiq®/desvenlafaxine
heroin	morphine
Vyvanse®/lisdexamfetamine[1]	d-amphetamine

---

[1] This is the correct generic spelling of the brand drug, Vyvanse®. This unique generic spelling is likely a play on the word "amphetamine," which is a contraction from alpha-methylphenethylamine.

The reasons behind a prodrug's development are varied. It may be a necessity due to problems with bioavailability and poor absorption. It may be an unstable molecule that requires the liver's help to stabilize. Difficulties in formulation and adverse effects of toxicity may be behind the development. Finally, it may be the need for capital gains. A prodrug can be sold first. When the patent expires and becomes an eligible generic, the original drug manufacturer can introduce a "new drug" which is the active metabolite to begin with along with a new patent life and a much more expensive brand name price.

Research and developmental reasons aside, if a drug is a prodrug, and the patient has a healthy liver, then no problem. If a drug is a prodrug, and the patient has liver difficulties, then prodrugs should be avoided. A healthy, functioning liver is required to activate an inactive compound into an active drug moiety – one that will heal where needed. If a person's liver is compromised, then the drug will not be metabolized into an active compound. If this is the case, an inactive drug will enter systemic circulation and do nothing.

Bottom line. Liver functioning properly? Prodrug will work fine. Liver impaired? Avoid prodrugs and pick a drug that does not tax an impaired liver.

QUIZ TIME

(1) Which of the following is considered the prototypic prodrug?

     (a) aspirin

     (b) codeine

     (c) Thorazine®

     (d) Risperdal®

(2) A prodrug is a _____, something that comes before.

     (a) novelty

     (b) postcursor

     (c) cursor

     (d) precursor

(3) If a patient has a compromised _____, avoid prodrugs.

     (a) lungs

     (b) kidney

     (c) brain

     (d) liver

(4) A prodrug is a biologically inactive compound that is activated by the _____ and cleared by the _____.

     (a) lungs; nasal passages

     (b) heart; venous system

     (c) liver; kidneys

     (d) kidneys; liver

(5) Rank the order from 1 to 4 (with 1 being first and 4 being last) in which drugs receive their names.

     _____Generic

     _____Brand

     _____Street

     _____Chemical

ANSWERS

(1) The correct answer is "b," codeine. In the liver, codeine is broken down by uridine diphosphate (UDP) glucuronsyltransferase (Phase II) into 5% — 10% morphine. If the patient has a healthy liver, this is an acceptable choice.

(2) The correct answer is "d." A precursor is required for the creation of a new entity.

(3) The correct answer is "d." The liver is the organ that holds the majority of the responsibility for breaking down, biotransforming, inactivating, and metabolizing prodrugs.

(4) The correct answer is "c." The liver activates the prodrug; and as with any other drug, the kidneys clear the prodrug's metabolite from the system.

(5) Drugs are named in the following order: (1) chemical, (2) generic, (3) brand, (4) street. Of the four, slang street names are ever changing; but chemical, generic, and brand remain constant.

Module 11

Love is Blind

Inducers vs Inhibitors

☺

It takes a mental stretch to think of the liver as a sexy organ. However, as it is in any intimate encounter (assuming it is possible to anthropomorphize this large, complicated, reddish-brown glandular organ), the liver can be stimulated, seduced, and excited to overrespond; or the liver can spurn advances from a drug suitor and grant less affectionate gestures. As funny as it sounds, that is exactly what happens with some drugs and CYP450 enzymes.

Some drugs can excite the liver to overrespond and create excess enzyme; or some drugs can inhibit the liver and produce less CYP450 enzyme. If a drug "excites" the liver, and more CYP450 enzymes are produced, that drug is an inducer. Taking the opposite approach, in this "sexy" example, a drug can "turn off" a liver and reduces its production of CYP enzymes. In that case, this drug would be an inhibitor. Inducers cause the liver to produce more CYP enzymes; inhibitors cause the liver to decrease its production of CYP enzymes.

Inducers increase enzymes; inhibitors decrease enzymes; and enzymes break down substrates. Three essential words: (1) substrate – a new term, (2) inducers, and (3) inhibitors.

There is a relationship between inducers/inhibitors and their substrate (from Latin, "substratum" meaning to "spread out, lay down, stretch out, or give in to"). A substrate is any medication that is broken down or metabolized by the enzymatic system. Think of it this way: A substrate is the "victim" of an enzyme's action. The enzyme will break down its target or substrate. For instance, if one reads that "Elavil®/amitriptyline is a substrate of the 3A4 (along with 2D6 and 2C19) enzyme," that means Elavil®/amitriptyline (the substrate) is broken down (victimized) by the CYP450 3A4 enzyme. In the game rock, paper, scissors, if all three are substrates of the CYP450 enzymes, then enzymes rule and can break down rock, paper, and scissors.

To put all this together, try the following fictional (which can happen and has happened) example.

Soltamox®/tamoxifen is not a psychotropic medication but is a hormone receptor-positive breast cancer medication. It is also a prodrug. When swallowed, inactive tamoxifen moves from mouth to throat, throat to stomach, stomach to small intestine, then via the hepatic portal vein, inactive tamoxifen finds its way to the liver. Soltamox®/tamoxifen is a substrate (a victim or a target) of numerous CYP450 enzymes including 2B6, 2C9, 2C19, 2D6, and 3A4. Translated: CYP enzymes break down tamoxifen into an active metabolite (the perfect definition of a prodrug). This active metabolite, endoxifen, now enters the system and heads to the job site to treat the patient's breast cancer. Putting this together: Soltamox®/tamoxifen is a prodrug that is a substrate of CYP450 enzymes. These enzymes break down the parent drug, Soltamox®/tamoxifen, into an active metabolite called endoxifen, which serves the patient's needs and treats the patient's breast cancer.

No doubt any patient dealing with breast cancer will experience a continuum of moods, from sadness to full-blown depression. Imagine a patient, currently taking tamoxifen, is also prescribed what is thought to be a "benign" antidepressant like Prozac®/fluoxetine. At face value, it sounds as if the prescriber is doing their best to remove the sting of cancer diagnosis and to help the patient cope. However, there is a problem. The antidepressant and anticancer medications do not mix and can reduce the effectiveness of tamoxifen (Bezerra, et al., 2018).

Prozac®/fluoxetine is a CYP450 2D6 enzymatic inhibitor. That means this "benign" antidepressant influences the liver. The result is Prozac®/fluoxetine enzymatic inhibition. Now, the liver makes less than normal amounts of the 2D6 enzyme. The result is less enzyme than normally required to break down the inactive product, Soltamox®/tamoxifen, into active endoxifen. As a result, much of the original medication remains unchanged, stays in its inactive form, and circulates the body doing nothing to control the disease.

If there is an insufficient enzyme, for whatever reason, to biotransform the substrate (prodrug, in this case), the active metabolite will not emerge. The drug is rendered useless and is of no medicinal value to the sufferer.

A prodrug needs the normal amount of enzyme to form the active metabolite. The antidepressant stopped this breakdown because Prozac®/fluoxetine reduces, inhibits the liver's typical output of enzyme. Not enough enzyme, the drug is not broken down properly. The result: Medication error.

When regular drugs (not prodrugs) are mixed incorrectly with other drugs, similar enzymatic inducing or inhibiting actions are frequent. Imagine "Drug A," all alone, taken enterally, by mouth. This drug is a substrate of one of the CYP450 enzymes, and the liver is functioning correctly, making the right amount of enzyme necessary to metabolize "Drug A." Not a problem. All is right in this physiological world.

Now imagine being prescribed "Drug B," which is to be taken along with "Drug A." This new drug affects the liver's production of CYP450 enzymes. "Drug B" is an inducer of the enzymes that metabolize "Drug A." Inducers cause the liver to make more enzymes. Now, mixing "Drug B" (the inducer) with "Drug A", enzymes are increased and less of "Drug A" will be available to make its way to the systemic circulation. In effect, the dose is reduced because extra CYP enzymes break down or inactivate even more of "Drug A."

Here is an example. Many psychotic patients smoke cigarettes. This is an example of the dose of an antipsychotic drug being affected by cigarette smoking, which is another "medication of sorts." Smoking a cigarette is equivalent to adding a CYP450 1A2 inducer. Smoking induces liver enzymes, which create more enzymes to break down more of the antipsychotic medication. The result, less antipsychotic drug finds its way to the systemic circulation. This is equivalent to an accidental reduction in dose.

Imagine "Drug C," all alone, taken enterally, by mouth. This drug is a substrate of CYP450 enzymes, and the liver is functioning properly making the right amount of enzyme necessary to metabolize "Drug C." Not a problem. Once again, all is right in this physiological world.

Now imagine being prescribed "Drug D," which is to be taken along with "Drug C." This new drug affects the liver's production of CYP450 enzymes. "Drug D" is an inhibitor of enzymes that metabolize "Drug C." Inhibitors cause the liver to make less enzyme. Mixing "Drug C" with "Drug D" (the inhibitor), enzymes are reduced and more of "Drug C" makes its way to the systemic circulation.

Here is another real-world example. Imagine a patient needs 100 mg of "Drug C," and has heard there is a way to save money on medications. The patient has heard that if grapefruit juice (a CYP450 3A4 inhibitor) is taken with "Drug C" the dose of the medication will be increased. In this case, the patient cuts the medication in half (50 mg) and drinks five ounces of grapefruit juice along with the medication. Mixing a CYP inhibitor (grapefruit juice) with the medication indirectly increases the dose (50 mg) of her drug. Drug plus an enzymatic inhibitor drug will increase the dose. One important disclaimer: This is dangerous and is NOT advised without consulting an individual's medical care advisor.

CYP450 enzymes, substrates, inducers, inhibitors. Anybody can prescribe. Doing it correctly takes talent.

QUIZ TIME

(1) Imagine a patient who is taking a certain form of birth control that is a substrate of a CYP450 enzyme. The patient mixes another prescribed medication, one that is a CYP450 inducer of the same enzyme, with the contraceptive. What will happen to the effectiveness of the contraceptive?

  (a) nothing

  (b) less contraception will find its way into circulation

  (c) more contraception will find its way into circulation

  (d) the pharmacist and the prescriber will be sued.

(2) A substrate is _____ of an enzyme.

  (a) another word for medication

  (b) another name for CYP enzymes

  (c) an inducer

  (d) a target of an enzyme

(3) True or False? No CYP450 2D6 inducers have been discovered to date.

(4) True or False? Medication errors can never be eradicated entirely, but they can be reduced.

(5) Another name for Phase I Metabolism is _____.

  (a) conjugation

  (b) functionalization

  (c) glucuronidation

  (d) acetylation

ANSWERS

(1) The correct answer is "b." Adding a prescribed inducer will metabolize and break down more birth control medication, the result, pregnancy. The case can be made that "d" is the correct answer. There are cases on record where a prescriber mixed an inducer with a contraceptive, pregnancy ensued; and the prescriber and the dispensing pharmacist were held legally responsible.

(2) The correct answer is "d," a substrate is a target of an enzyme.

(3) TRUE. You are correct, this was not covered in this module. The question's purpose is to provide new information and to explain that the field of psychopharmacology is fraught with peculiarities and exceptions. Psychopharmacology done correctly is a woefully humbling task. So, give yourself credit if you answered either "True" or "False," but remember the correct answer. No CYP450 2D6 inducers have been discovered to date. Give it time.

(4) TRUE. Medication errors will happen as long as healthcare providers prescribe, patients take, and caregivers give them. But they can be minimized. This question is designed to point out the importance of a basic understanding of CYP450 enzymatic action. This is an easy place for drug mistakes to hide.

(5) The correct answer is "b," functionalization. The other three answers, "a," "c," and "d," pertain to Phase II, conjugation.

Enemy at the Gate

Pharmacodynamics

☺ ☺ ☺

Drugs (xenobiotics) and the body are not really enemies – although that is how your liver is programmed to treat them. The liver responds as if everything, including foods, are also enemies until proven benign. Drugs are neither good nor bad; they just are. How they are used determines the quality of good or bad. As Paracelsus (c. 1493-1541, real name was Theophrastus Phillipus Aureolus Bombastus, but being the narcissistic "rock star" expert of his time, he changed his name to "Paracelsus"), the "Father of Pharmacology and Toxicology," stated, "Everything is poison. It is the dose that determines whether it kills or cures." Being a man of letters, and a bit of a short-statured snob, he stated the above in Latin: "Dois sola facit venenum," which translates, "The dose alone determines the toxicity."

Are drugs (xenobiotics) enemies? Occasionally yes, mostly no, and it always depends on the dose to determine the effect's intensity. Overall, it is contingent on pharmacodynamics (what the drug does to the body). But remember, drugs do not create any unique effects. They can only modulate, mimic, or block the actions of natural body chemicals.

Pharmacodynamics (from Greek, "pharmakon," and "dynamis" meaning "power") is what the drug does to the body. Pharmacodynamics is the drug's power over the body. Note that Pharmaco<u>d</u>ynamics has a "D" at the beginning of the fourth syllable. Drugs also begin with the letter "d." Pharmaco**D**ynamics is what the **D**rug does to the body.

Put another way, "Pharmacodynamics is the study of the physiological and biochemical interaction of drug molecules with the target tissue that is responsible for the ultimate effects of the drug" (Meyer & Quenzer, 2019, p. 27). A drug searches for its target — the same way a key interacts with a lock, a friend pursues a best friend, a lonely heart desires companionship, a parasite needs a host, a postal worker finds a mailbox, a ship's movement requires a semaphore, or a prizefighter needs a knockout.

To understand pharmacodynamics – what the drug does to the body – a cursory understanding of (1) receptors, (2) ligands, (3) agonists, and (4) antagonists is the perfect place to start.

One word of warning: In the biochemical and pharmacological world, nothing ever has one name. They will always have several names. Admittedly there are nuanced differences; however, they will have common sounding and esoteric names. Most make sense; some do not.

It all starts at the receptor, usually a large protein (which is a complex chain of various amino acids) on a cell's membrane between polar heads and lipid tails, within cells, and possibly other places not yet discovered. Referred to as receptor proteins and shaped like little boulders, they pick up signals and pass messages between cells. There are four major ways (keys to understanding psychotropic drug action) they can signal action. Receptors can signal action through (1) ion channels, (2) G proteins, (3) carrier proteins, and (4) enzymes.

(1) Ion Channel Receptors (aka Rapid Ligand Gated Ion Channels). These receptors are easy to identify. They have a central pore that, with appropriate stimulation, can open and close. This central pore forms an "ion" (an atom or molecule with a net electrical charge) channel. When it opens, it allows the flow of a specific ion (e.g., hydrogen, H^+, and chloride, Cl^-). through this enlarged pore. This is an amazingly fast receptor. It responds quickly to stimulation. One of the most well-known is the GABA-A (GABA is a neurotransmitter, and "A" is the specific name of the individual receptor) Chloride Ionophore. This receptor is also called the "GABA-A receptor," the "GABA-benzodiazepine receptor complex," or "GABA receptor-chloride ion channel complex."

Originally called "ion complexers," David Pressman, in 1965, refined and coined the term "ionophore." "Ion" comes from Greek, "ienai," meaning "to go"; and "phore" derives from Greek, "phero" meaning "carrier." Put the two together and "ionophore" means "to go carrier." In 1834 polymath William Whewell, selected the term, "ion" to represent the unknown "something" that "goes" from an electrode through an aqueous solution. This is what these ionophores accomplish. The pore opens and ions are carried through into the cell. These ligand gated receptors are also called "ionotropic," which just means a cell membrane that opens or closes in the presence of ions.

(2) G Protein-Coupled Receptors (GPCR; aka Rapid G Protein-Coupled Receptor System). These receptors are a large and diverse family of proteins. When these receptors are activated, they release an intracellular protein (a G protein, short for "guanine nucleotide-binding protein") that controls enzymatic functions. It is not a pore or an opening and does not allow direct passage of ions – at least not at first. Instead, it is a molecular switch that starts a cascade of action, like dominos falling one after the other. When it is stimulated, a G protein is activated and somewhere down the line, it opens or closes an ion channel at another place on the cell membrane. The G protein is the middle-person, communicating between the initial receptor complex and intracellular enzymes (called "second messengers") to produce a response. Examples of second messengers are cAMP, cGMP, inositol triphosphate (IP_3), diacylglycerol, and calcium. They are metabotropic because these receptors are linked to metabolic processes that require energy for a response. These G protein-coupled receptors are fast but not as fast as ionotropic receptors.

(3) Carrier Proteins. Also known as "Transport Proteins," these receptors are located in the membranes of neurons and can bind to specific neurotransmitters in the synaptic cleft and transport them to the source that released them. Working against the "concentration gradient," they are shaped like little bowls that trap and cradle the molecule and return it to its source. It is a one-way revolving door that traps and then dumps on the other side. Neurotransmitters are repackaged and released later. One-third of all psychotropic medications manipulate this system by blocking the return or reuptake of neurotransmitters via the transporter. Serotonin

reuptake inhibitors block "carrier proteins," so the serotonin molecule, a neurotransmitter, is blocked and cannot return home.

(4) Enzymes. The fourth type of receptor protein or message pathway that regulates drug action is enzymatic and facilitates the breakdown and regulates neurotransmitters' availability. Easy to spot, as a rule, they end in "-ase." So, acetylcholinesterASE facilitates acetylcholine breakdown; monoamine oxidASE breaks down the monoamines like dopamine, norepinephrine, and serotonin.

To summarize, what do receptors do? They receive messages, respond, and pass the signal down the line. Where are they? Either on the cell or inside the cell. How fast are they? Very. Ligand-gated, ionotropic receptors are the fastest; intracellular receptors are slower but still fast. Why are receptors important in psychopharmacology? They can be the site of drug action.

Drug action is categorized by the body action it triggers. Overall, there are four types of responses. Drugs can (1) stimulate or block cellular activity, (2) replace essential body compounds (e.g., insulin), (3) interfere or inhibit growth as it does with infectious bacteria, or (4) irritate cells and cause a natural response. For instance, laxatives irritate the colon wall and increase movement (Kamienski & Keogh, 2006).

It is important to know that a receptor will only do what it is biologically "wired" to do naturally. This is called "intrinsic activity," the drug-receptor complex's ability to produce a maximum functional response. In other words, when the ligand stimulates the receptor, it responds exactly as the receptor is programmed. No matter how much a ligand "begs, cajoles, teases, or pleads," the receptor will only do what it is intrinsically wired to do. That is a good thing and explains why off-label prescribing does not work well. Most off-label medications are prescribed for their side-effect profile.

Receptors patiently wait ("resting state") for something to attach to them. This binding only happens when there is affinity (from Latin, "affinitatem," meaning "relationship by marriage"). If there is affinity, a marriage of sorts, something is wed or bound to the receptor, and the receptor responds. This is where ligands come into play. A ligand is a tie that binds. It is the "attacher" to the "attachee" (receptor).

What can attach to a receptor? A ligand (from Latin, "ligandus," meaning "to bind or to attach"). A ligand is anything that attaches, connects, fastens, joins, and/or links. As the tie that binds, it is a union, nexus, or in modern parlance, a hookup. Anything that attaches to a receptor is technically a ligand.

Examples of ligands include lipoproteins, ions, immunoglobulins, neurotransmitters, drugs, and hormones. However, psychopharmacologists are more interested in two of the above — neurotransmitters and drugs.

The goal of psychopharmacology is to identify and to create medications that can attach with affinity (pretend to be something natural) and yield a clinically useful effect, help the patient, and protect them at the same time. Most of the time that happens; sometimes it goes awry.

Admittedly biochemical and pharmacological terms can be daunting, even overwhelming; fortunately, it is possible to understand "how all this works" with a modest background in this area. More about receptors in upcoming modules, but first a discussion of ligands, agonists, and antagonists, forthcoming in the next module.

QUIZ TIME

(1) For anything to bind to a receptor, there must be _____.

    (a) affinity

    (b) intrinsic activity

    (c) bioavailability

    (d) biotransformation

(2) Which of the following is a drug category of action?

    (a) stimulate

    (b) replace

    (c) inhibit

    (d) irritate

    (e) all the above

(3) _____ is what the body does to the drug.

    (a) pharmacodynamics

    (b) pharmacokinetics

    (c) pharmacoavailability

    (d) pharmacotransformation

(4) True or False? A ligand is anything that binds to a receptor.

(5) True or False? All risks associated with pharmacogenomics or pharmacogenetics have been identified.

ANSWERS

(1) The correct answer is "a." Affinity is what allows the key to be inserted in the receptor lock.

(2) The correct answer is "e." Drugs are categorized by their action in the body. They stimulate, block, replace, inhibit, and irritate the body's natural action.

(3) The correct answer is "b." Pharmacokinetics is what the body does to the drug. Pharmacodynamics – remember it has the letter "D" in it – is what the "D" for drug does to the body.

(4) TRUE. A ligand is a key to the lock.

(5) FALSE. Time will tell. New pharmacogenomic risks will emerge as the field continues to be defined and advances made.

A QUICK FYI. . . .

G protein-coupled receptors represent a relatively new field of study and demonstrate that the fields of pharmacology and psychopharmacology continue to positively evolve. What is known today pales in comparison to what will be the "future known." Second messenger receptors were not discovered until 1956 by Earl Wilbur Sutherland, Jr. T. W. Rall and he discovered cyclic AMP (cAMP), the almost universal "second messenger." The importance of this discovery cannot be minimized and is a significant discovery in molecular biology. As a result, it garnered these two esteemed scientists the 1971 Nobel Prize (physiology and medicine) and demonstrated the pervasiveness and prime importance of these compounds in all living things. Before this discovery, hormones were thought to be the sole regulatory substance in living organisms. Drs. Sutherland and Rall proved this view inaccurate and opened wide a pharmacological field that could now design medications to impact second messenger functioning and help numerous patients, especially those with cancer and diabetes (Dr. Earl W. Sutherland, Jr., Dies, 1974).

Module 13

The Bind That Ties

Ligands

☺ ☺ ☺ ☺

A ligand is anything that binds to a receptor. It is like the round peg looking for a round hole. If a ligand is a square peg, it just will not fit in a round hole. A receptor is picky. If it is not the perfect round shape, then that square ligand is rejected, rebuffed, and pushed away. It must be a perfect match for the neurotransmitter to bind to the receptor.

While there are several types of ligands, psychopharmacology focuses on neurotransmitters. They are chemical substances, released upon arrival of an action potential at the end of the nerve fiber, diffuse through the synapse (cleft or junction), and transfer an impulse to a receptor, nerve fiber, or muscle fiber. Pharmaceutical companies focus on the drugs that manipulate these neurotransmitters. Pharmaceutical companies make round drugs that fit in round receptors — or at least they try.

To date, approximately one hundred neurotransmitters have been identified; however, only a small fraction (some say twenty or less) of those neurotransmitter's actions are understood. When it comes to psychopharmacology, there are "The Big Seven" neurotransmitters. Most of the psychopharmacological medications prescribed will utilize one or more of these seven. They are (1) norepinephrine, (2) epinephrine,

(3) dopamine, (4) serotonin, (5) acetylcholine, (6) GABA, and (7) glutamate. Histamine should be included, but it is often underappreciated. Each of these neurotransmitters has its own module; but, walking before running requires understanding the neurotransmitter families, agonists, and antagonists.

These neurotransmitters can be grouped by type. Of these seven, four are monoamines, acetylcholine (it is its own family, in and of itself; it is that important; explanation to follow), and two amino acids, GABA and Glutamate (Glu).

First, Monoamines. Look at the word: "mono" and "amine." Monoamines are any biogenic (produced or brought about by living organisms) amine neurotransmitters with a single, mono, amine group on their chemical structure. No surprise that "dopamine" is a monoamine. The "amine" part fits the description of a monoamine.

Monoamines can be divided into two subtypes: catecholamines and indolamines. Of the big seven, three of these neurotransmitters fall into the catecholamine group. They are dopamine (DA), epinephrine (Epi), and norepinephrine (NE). Serotonin (5-HT, which translates to 5-hydroxytryptamine, and the "5" means it has a hydroxyl group in the 5th position) along with melatonin falls in the indolamine subtype. These are not the only neurotransmitters that are monoamines, but for psychopharmacology these are the most important.

Acetylcholine (ACh) is its own category. As one of the major seven neurotransmitters, it is incredibly important. Given its control over the parasympathetic, sympathetic, and somatic nervous systems, it could be argued that acetylcholine is the most important neurotransmitter in both pharmacology and

psychopharmacology. Perhaps the king of neurotransmitters.

MONOAMINES	
CATECHOLAMINES (Derived from Tyrosine)	INDOLAMINES (Derived from Tryptophan)
Dopamine (DA) Epinephrine (Epi) Norepinephrine (NE) *Students use the mnemonic "CATS live in the DEN" as way to remember the main three catecholamines. However, there are other catecholamines, including tyrosine and tyramine.	Serotonin (5-HT) Melatonin For the record: Histamine is neither a catecholamine nor indolamine. It is metabolized from the precursor, histidine.

Finally, GABA and glutamate are two amino acids. They are not monoamines;

they are amino acids. GABA is an abbreviation for "Gamma Aminobutyric Acid," and

glutamate (not an abbreviation but an anion of glutamic acid). For now, just

remember that GABA is inhibitory, and glutamate is excitatory.

In the field of psychopharmacology, these seven neurotransmitters are the

most important ligands. In fact, a substantial percentage of all psychotropic

medications are designed to manipulate any one or more of these seven

neurotransmitters.

QUIZ TIME

Match the Neurotransmitter to its Family/Class/Sub-type.

(1) _____ Indolamine

A. GABA

(2) _____ Catecholamine

B. Acetylcholine

(3) _____ Integral to Body's Nervous Systems

C. DA, 5-HT, & NE

(4) _____ Monoamines

D. Serotonin

(5) _____ Amino Acid

E. Epinephrine

(6) True or False? A ligand is anything that binds to a receptor.

(7) True or False? Serotonin is a catecholamine.

(8) True or False? GABA and Glutamate are monoamines.

(9) True or False? Acetylcholine is a monoamine.

(10) True or False? A large majority of drugs are made to manipulate the big seven neurotransmitters and facilitate receptor action.

ANSWERS

Matching: (1) D; (2) E; (3) B; (4) C; (5) A.

(6) TRUE. A ligand can be an ion, an immunoglobulin, a carbohydrate, a neurotransmitter, a drug, or hormone. If it attaches with affinity and precipitates a response, it is a ligand.

(7) FALSE. Serotonin is an indolamine.

(8) FALSE. GABA and glutamate are pure-bred, full-fledged amino acids.

(9) FALSE. Acetylcholine is special and is considered a "non-amine-amine." It technically is a quaternary amine.

(10) TRUE. These seven neurotransmitters represent foundational knowledge.

Play the "ASE"
Enzymes

☺ ☺ ☺

What happens when the body releases too many neurotransmitters? The two keywords in this question are "too many." When too many neurotransmitters are released and overwhelm the receptors, death can follow. This answer may sound like an exaggeration, but it is clear-cut. Many drugs are lethal in overdose because too many neurotransmitters are released; the body is overwhelmed, and organs cease functioning.

Fortunately, the body has a system to deal with this issue and preclude this from happening. Enzymes break materials down and render them harmless. Or to be technical, enzymes catalyze the metabolic processes that occur within a cell.

Initially discovered in the 1830s, the term was not coined until 1878. "Enzyme" derives from the Greek words meaning "with yeast" or "leavened." Because of yeast's reaction in fermentation (which speeds up the process, establishes predictability and control in the chemical reaction), it seemed the perfect simile – enzymes are like yeast.

There are six different classes of enzymes — all with polysyllable-hard-to-pronounce words, each categorized in terms of their specificity and beyond this module's scope. However, enzymes have an exclusive relationship with their substrates (the molecule to which an enzyme binds). Putting this concept together, all the monoamines have a specific enzyme (some have several enzymes) designed to break them (the substrates) down as quickly as possible.

Most enzymes' names can be divided into two parts. The first identifies the substrate upon which the enzyme attacks, and the second term defines the type of reaction. For instance, histamine-N-methyltransferase (HMT or HMNT) advises straight away that it is an enzyme (ends in "-ase"). The second part, "methyltransferase," explains how the enzyme does what it does. It methylates, transfers a methyl (CH_3) group to the substrate. Histamine is also broken down by diamine oxidase (DAO), found mostly in the kidneys, intestinal lining, and thymus.

Usually, enzymes are easy to spot because many of them end in "-ase." However, there are common enzymes out there that do not end in "-ase." For instance, pepsin, trypsin, and chrymotrypsin are all common enzymes that break down substrates in the digestive system, but they do not end in "-ase." However, they are enzymes nonetheless (Beck, 2019).

The previous module identified "monoamines." What breaks monoamines down? Monoamine oxidase (MAO). It is no stretch of knowledge to figure out that the "catecholamines" are broken down by catechol-o-methyltransferase. Acetylcholine (ACh), a unique non-amine-amine, is broken down by acetylcholinesterase (AChE), which breaks down 25,000 molecules of acetylcholine per second by hydrolysis in one easy step.

Remember the earlier question, "What happens when too many neurotransmitters are released?" The body has a mechanism to reduce the probability of this happening. Enzymes stop neurotransmitters by breaking them down quickly and indirectly suppressing a nerve's excitation or the transmission of an impulse. Enzymes are one of the body's fail-safe mechanisms to keep one functioning.

Putting this together, dopamine, epinephrine, and norepinephrine – monoamines and catecholamines – are broken down by monoamine oxidase (MAO) and catechol-O-methyltransferase (COMT). Most clinicians use abbreviations. Serotonin is broken down by monoamine oxidase (MAO). There is no such thing as an "Indolaminase." Histamine is broken down by histamine-N-methyltransferase (HMT, HNMT) and diamine oxidase (DAO) (Beck, 2019).

ENZYMES AND SUBSTRATES	
ENZYME Monoamine Oxidase (MAO) and Catechol-O-Methyltransferase (COMT)[1]	ENZYME Monoamine Oxidase (MAO)
MAO & COMT SUBSTRATES Catecholamines Dopamine (DA) Epinephrine (Epi) Norepinephrine (NE)	MAO SUBSTRATE Indolamines Serotonin (5-HT)

QUIZ TIME

Match the following to the enzyme that breaks it down. Some enzymes can be used more than once, and several have more than one answer.

(1) _____ Serotonin (5-HT)   A. Acetylcholinesterase (AChE)

(2) _____ Epinephrine (Epi)   B. Monoamine Oxidase (MAO)

(3) _____ Histamine (H)   C. Catechol-O-Methyltransferase (COMT)

(4) _____ Acetylcholine (ACh)   D. Histamine-N-Methyltransferase (HMT/HMNT)

(5) _____ Norepinephrine (NE)   E. Diamine Oxidase (DAO)

(6) _____ Dopamine (DA)

ANSWERS

(1) The answer is "B." Serotonin is broken down by Monoamine Oxidase (MAO); it is not a catecholamine. It is an indolamine, so MAO does the trick.

(2) The answers are "B" and "C." Epinephrine is both monoamine and catecholamine. Both MAO and COMT break it down.

(3) The answers are "D" and "E." HMT/HMNT and DAO.

(4) The answer is "A." Acetylcholine is unique, necessary; and has its own personal enzyme: Acetylcholinesterase (AChE).

(5) The answers are the same for epinephrine, "B" and "C." Norepinephrine is both a monoamine and a catecholamine. It is broken down by MAO and COMT.

(6) The answers are "B" and "C." Dopamine is both monoamine and catecholamine. MAO and COMT break it down.

Module 15

Let the Games Begin
Agonists vs Antagonists

☺ ☺ ☺ ☺

Receptors. Covered.

Ligands. Covered.

Two remaining concepts: Agonist and antagonist.

These are not difficult terms to master, and they are an integral characteristic of every ligand, neurotransmitter, or drug. Since nothing in psychopharmacology is as simple as it sounds, there are agonists, antagonists, partial agonists, inverse agonists, and indirect agonists. Those terms are worthy of a basic understanding, and the discussion follows. However, there are even more terms to describe ligand-receptor action: co-agonist, selective agonists, competitive antagonists, reversible and irreversible agonists, mixed agonist/antagonist, and allosteric modulators. All said, the two basic concepts, agonist and antagonist, is the place to start.

It goes back to the original Greek games – the Olympics. Agonist comes from Greek "agonists" and means "contestant." So, "agon" is the game, and "agonia" is the struggle for victory, specifically mental struggles. The true meaning of "agon" and

93

"agonia" roughly translates, "The harder it is to achieve, the more worthy the task." One more etymology for fun: "agonizesthai" means "to contend" for a prize.

If the agonist is the contestant, then the antagonist is the opponent, competitor, villain, enemy, rival, or blocker.

The term "agonist" was first used in 1658, and its definition really has not changed. An agonist is a chemical – drug, neurotransmitter, hormone – that binds with affinity to a receptor and activates a prewired, biological response. If the drug is an agonist and binds with affinity, it mimics a natural endogenous (originates in the body) chemical and triggers intrinsic activity. Suppose a drug is a round peg, and the receptor is a round hole. In that case, the result is a beautiful relationship between neurotransmitter or drug and receptor binding plus a conformational change.

Conformational change means that the receptor changes its shape. It is like a receptor action that acknowledges the will of the ligand (drug or neurotransmitter). One other easy point. If the agonist elicits a full, maximum response from the receptor, then this is a "full agonist."

Sometimes a round peg has an affinity (binds to the receptor) and is more of a "plug" than an activator. This is an antagonist. It blocks any other ligand from binding and agonizing the receptor. It is a cork, stopper, lid, cap, or receptor cover. Perhaps the most used term for antagonist is blocker. When an antagonist binds and blocks, agonists may be in the vicinity and cannot bind since that receptor is already engaged. The antagonist got there first. The cork keeps things out; the receptor remains in its resting state, and no intrinsic action occurs. Probably the two most famous psychotropic, pharmacological drugs of all, Thorazine®/chlorpromazine and Haldol®/haloperidol (affectionately referred to as "Vitamin H") are antagonists; and to be specific, technically they are dopamine two (D₂) antagonists. Translated: These drugs block the "dopamine receptor, number two."

A partial agonist is complicated; it can have both agonist and antagonist qualities. Technically, it is an agonist that fails to produce maximal activation in an entire receptor group – no matter how much of the drug is applied. However, if a partial agonist and a full agonist have an affinity for the same receptor, they will compete. In that case, the full agonist's ability to produce its maximum effect is reduced. Perhaps it is best to think of partial agonists as modulators. Standard psychotropic partial agonists include Buspar®/buspirone (antianxiety medication) and Abilify®/aripiprazole (a multipurpose drug used to treat psychoses, bipolar disorder, and agitation in autistic children).

Approximately fifty percent of all medications function as either agonists or antagonists.

Now for the tricky one. Most students are taught, or accidentally led by default, that the opposite of an agonist is an antagonist. Incorrect. To explain, an agonist mimics receptor action creating an intrinsic activity or a specific prewired response. The agonist causes the receptor to act. An agonist causes something to happen. An antagonist, on the other hand, is a blocker, cork, or stopper. While an antagonist binds with affinity, nothing happens. An antagonist, by definition, keeps the receptor from responding to anything. An agonist gets a response; an antagonist allows the receptor to do nothing but remain in a resting state. Therefore, an antagonist cannot possibly be the opposite of an agonist.

The opposite of an agonist is an "inverse agonist." It binds; it causes the receptor to do the opposite of what it is intrinsically wired to do. Examples of inverse agonists include Benadryl®/diphenhydramine, Risperdal®/risperidone (for bipolar and irritability for autistic children), and Nuplazid®/pimavanserin (for Parkinson's disease associated psychosis). While the FDA has classified Nuplazid®/pimavanserin as an inverse agonist, there is debate concerning this distinction. Many believe it is more of an antagonist than an inverse agonist (Nutt & Blier, 2017).

Indirect agonists (not to be confused with inverse agonists) are sometimes called indirect-acting agonists. They are agonists but not traditional ones. They have intrinsic activity; they enhance the release or action of the body's natural neurotransmitters "indirectly." Indirect agonists have NO specific agonist activity at the receptor site but can cause an increase in the release of neurotransmitters or stop enzymes from metabolizing neurotransmitters. To put it another way, if a drug is an agonist, it is a straightforward round peg that binds to a receptor's round shape. However, an indirect agonist drug is sneaky. It indirectly increases neurotransmitter release which then agonizes receptors. There are three ways to indirectly increase the number of neurotransmitters, which then bind to receptors. An indirect agonist can (1) block transporters, (2) provoke neurotransmitter release, or (3) prevent neurotransmitter breakdown. Each of these can increase available neurotransmitters to bind with receptors.

(1) Block Transporters. Recall that the body will never allow neurotransmitters to hang around for no reason. Too many neurotransmitters released can be harmful to the body. The transporter, or reuptake apparatus, is a "door" that allows neurotransmitters to return to the terminal button for repackaging and rerelease. If this door is blocked, neurotransmitters are forced to hang around longer, thereby increasing the probability that each neurotransmitter will bind to a receptor. This is a classic indirect agonist. The drug, which blocks neurotransmitters' recycling, reuptake, or return home, indirectly increases the chances a neurotransmitter will agonize a receptor. Prozac®/fluoxetine (a SRI) is the textbook example. It blocks the door and allows no entry; as a result, serotonin molecules cannot return home and must do their job. Any drug that blocks transporters, called "reuptake blockers" or "reuptake inhibitors," is an indirect agonist — more on this in a future module.

(2) Provoke Neurotransmitter Release. Amphetamines are a classic example. By indirect action, they provoke neurotransmitter release. The result is an increase in neurotransmitters that can bind to receptors. Another example is 3,4-methylenedioxy-methamphetamine (sound it out and pronounce each syllable equally, MDMA) better known as Ecstasy. The mechanism of action (MOA) of MDMA is "SNDRA," serotonin norepinephrine dopamine releasing agent. The result in both cases is the same: An indirect agonist causes an increase of neurotransmitters to bind with receptors.

(3) Prevent Neurotransmitter Breakdown. Some drugs block the enzymes which break down neurotransmitters. By blocking the breakdown, they increase the number of neurotransmitters, and earn the moniker "indirect agonist." One of the best

examples of this is Aricept®/donepezil – a commonly used medication for dementia. It blocks the enzyme that breaks down acetylcholine (AChE; acetylcholinesterase), indirectly causing an increase in acetylcholine and indirectly agonizing more receptors. The technical mechanism of action (MOA) for Aricept®/donepezil is acetylcholinesterase inhibitor (AChEI).

It does not matter if it is an agonist, antagonist, inverse agonist, partial agonist, or indirect agonist. They all have one thing in common. Each describes how a neurotransmitter or medication can bind to a receptor, but each can vary in their ability and strength.

QUIZ TIME

(1) The opposite of an agonist is a(n) _____.

    (a) super agonist

    (b) antagonist

    (c) indirect agonist

    (d) allosteric modulator

    (e) none of the above

(2) Approximately _____ % of medications are either agonists or antagonists.

    (a) 50%

    (b) 30%

    (c) 0%

    (d) 100%

(3) Which of the following is a drug category of action?

    (a) stimulate

    (b) replace

    (c) inhibit

    (d) irritate

    (e) all the above

(4) Indirect agonists increase neurotransmitters by _____.

    a. blocking transporters

    b. provoking release

    c. blocking enzymes which breakdown neurotransmitters

    d. all the above

(5) Another word for antagonist is _____.

    (a) agitant

    (b) esteraser

    (c) inhibitor

    (d) affinity

ANSWERS

(1) The correct answer is "e." The opposite of an agonist is an inverse agonist. It binds with affinity and precipitates an opposite response.

(2) The correct answer is "a." Approximately 50% of all medications in the entire pharmacopeia are either agonists or antagonists.

(3) The correct answer is "e." Drugs are categorized by their action in the body. They stimulate, replace, inhibit, and irritate the body's natural action.

(4) The correct answer is "d," all the above. To increase neurotransmitters indirectly, they block, provoke, and inhibit. All in the name of increasing neurotransmitters. More neurotransmitters offer more opportunity to bind with receptors.

(5) The correct answer is "c," inhibitor, although "blocker" is more commonly used.

Module 16

The Perfect Triad

Potency, Efficacy, and Therapeutic Index (TI)

☺ ☺ ☺ ☺

When it comes to drug agonists, there are three important straightforward concepts: Potency, efficacy, and therapeutic index. Potency refers to strength; efficacy refers to usefulness/effectiveness; and therapeutic index refers to safety.

Potency asks, "How strong is the drug?" Potency refers to the strength of the medication or the number of drug molecules required to produce a biological response at the receptor site. Since most pharmacology is written "backwards," graphing potency on a dose-response curve (with milligrams on the horizontal axis and response on the vertical axis) reveals the lower the dose the more potent the medication. As the dose increases, the potency of the drug decreases. To explain, methamphetamine requires a lower amount to achieve the same effect as a "regular" amphetamine. Therefore, methamphetamine's dose is more potent.

Efficacy asks, "Does the drug work?" If it is a pain reliever, does the pain go away when the medication is taken? To put it another way, efficacy is the ability to produce the desired effect. If a drug has 100% efficacy, then it produces a greater response than the desired effect. Back to the dose-response curve, and to repeat, potency is graphed in milligrams on the horizontal axis; efficacy is graphed on the vertical axis from 0% to 100% for maximal effects. Concerning psychotropics, most are not used to their maximum effect because of toxicities and patients' inability to tolerate numerous and intense side effects.

The following chart evaluates the potency and efficacy of three medications: Drug A, Drug B, and Drug C. Drug A is more potent than Drug B which is more potent than Drug C. As the drug's dosage moves from left to right on the horizontal axis, it becomes less potent. Drugs A, B, and C are equally efficacious. On the graph, these three drugs have virtually the same effect. While Drug C has a higher dosage than Drug B or Drug A, all three achieve a maximal effect, which is roughly the same. Drug C is far less potent but just as efficacious as Drugs A and B.

The dose-response curve is also a helpful standardized way to graph receptor activity. It depicts "the extent of biological or behavioral effect (mean response in a population) produced by a given drug concentration or dose" (Meyer & Quenzer, 2019, p. 31). Technical terms aside, the dose-response curve is a picture or graphic representation of the relationship between potency and efficacy (strength and usefulness/response). The dose-response curve shows how little or how much it takes to get the job done. How little or how much (potency) does it take to lower blood pressure, provide pain relief, or reduce depression (efficacy). The graph provides a means of comparison. Other practical considerations that are just as important include side effect possibilities, cost, and frequency of dosage.

Efficacy and potency comparisons on the dose-response curve are usually made at the ED_{50} and TD_{50} (also called LD_{50}) levels. ED_{50} is the effective dose fifty, or the dose that produces a quantal (all or nothing) effect in 50% of the population that takes the medication. The ED_{100} would be the dose that has the desired impact in 100% of the population. The TD_{50}, also known as the toxic dose fifty (or LD_{50}, lethal dose fifty), is the dose at which 50% of the population experiences toxic, lethal effects.

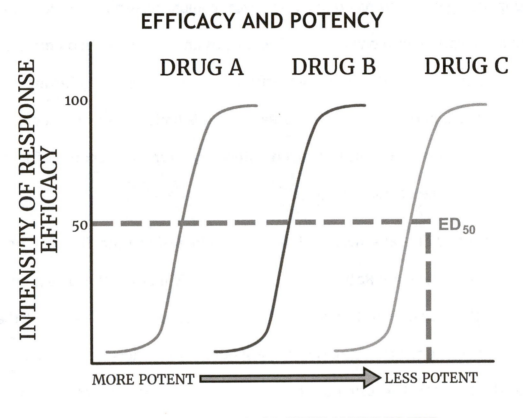

EFFICACY AND POTENCY

NOTE: AS DOSE RESPONSE CURVE MOVES TO THE LEFT, CLOSER TO "Y–AXIS," POTENCY INCREASES.

To explain the value of the dose-response curve, answer the following question. Which of the three drugs — hydromorphone (Dilaudid®), morphine, and codeine – is the best? Tough question. Because "best" has so many connotations and permutations, it is a difficult question to answer. How do you define best?

Pharmacologists define "best" as the most efficacious. As a result, it is the depiction of the ED_{50}, graphed on the dose-response curve, which confirms best. Of these three drugs, their potencies are different. Dilaudid®/hydromorphone (Drug A) is more potent than morphine (Drug B) which is more potent than codeine (Drug C). However, at the ED_{50} level, they are equally efficacious. If you were to throw aspirin into that mix (not an opioid), aspirin is far less potent than these three and far less efficacious than pain-relieving opioids. However, comparing aspirin to these opioids is like comparing apples with one orange. The opioids and aspirin work on different receptors. The three opioids are equally efficacious at the ED_{50} for relieving pain, but they are not equal in their potencies. Potency's use is limited; but efficacy is essential. When there is a choice between potency and efficacy, choose efficacy. A drug that works is the deciding factor.

The ED_{50} and TD_{50} are important variables in identifying the Therapeutic Index (TI; also called Therapeutic Ratio, TR). The Therapeutic Index (TI) is an estimate of the safety of the medication. It is a type of dose-response curve but looks at the percentage of the population experiencing a specific drug-induced response. It compares the ED_{50} and TD_{50} using the following ratio/formula: Therapeutic Index = TD_{50}/ED_{50}.

If the TD_{50} for a drug is 450 mg and the ED_{50} is 50 mg, then the therapeutic index would be 9 (TI = 9). This drug would require monitoring; it may not be safe. If a drug has a Therapeutic Index of at least 100, it is considered to have a "wide therapeutic index." That means that the dose that works is substantially less than the dose that causes toxic, lethal side effects. If the Therapeutic Index (TI) is below 100 then the drug is said to have a "narrow therapeutic index" — the dose that works is too close to the amount that causes substantial toxic side effects. For example,

Eskalith®/lithium has a very narrow therapeutic index — the Eskalith®/lithium dose that is effective is remarkably close to the toxic dose. Other drugs with a narrow therapeutic index include the Tri-Cyclic Antidepressants (TCAs), Lanoxin®/digoxin, and Coumadin®/warfarin. Fortunately, most antidepressants and antipsychotics have a wide therapeutic index.

Pharmaceutical companies' television commercials regularly promote the potency of their new and improved drugs. It is not potency that counts; it is efficacy. Efficacy always surpasses the power of potency. Maybe the drug company should market their efficacy and therapeutic index instead. If given a choice, either potency or efficacy, pick efficacy. What good is a potent drug that does not work?

QUIZ TIME

(1) Potency is the _____ of the drug; efficacy is the _____.

      (a) strength; response

      (b) response; strength

      (c) ED_{50}; TD_{50}

      (d) TD_{50}; ED_{50}

(2) The opposite of an agonist is a(n) _____.

      (a) super agonist

      (b) competitive agonist

      (c) indirect agonist

      (d) inverse agonist

(3) LD_{50} and TD_{50} are _____ terms.

      (a) equivalent

      (b) opposite

      (c) dissimilar

      (d) contrasting

(4) Which of the following drugs does NOT have a narrow therapeutic index?

      (a) Eskalith®/lithium

      (b) Lanoxin®/digoxin

      (c) Coumadin®/warfarin

      (d) Prozac®/fluoxetine

(5) On the dose-response curve, as the dose amount increases and moves to the right on the horizontal axis, the potency of the drug _____.

      (a) stays the same

      (b) decreases

      (c) increases

      (d) fluctuates based on efficacy

BONUS QUESTION

Your next door neighbor has heard that you are studying psychopharmacology, and she wants to know which drug is better for "racing thoughts" – Fixafun, Coszpain, or Happarx (these are all fake drugs, not real, made up; for educational purposes only). You advise that you will check it out for them at no charge. After careful consideration, you decide the correct way to determine the "better drug" is to graph, evaluate, and compare each drug's efficacy, one to another. From the choices below, which measuring device will you use to assess and to determine which is "the best," most efficacious drug?

(a) CYP_{50}

(b) RX_{50}

(c) TI_{50}

(d ED_{50}

ANSWERS

(1) The correct answer is "a." Potency is the strength of the medication; efficacy is the response.

(2) The correct answer is "d." An inverse agonist is the opposite of an agonist.

(3) The correct answer is "a." Both LD_{50} (Lethal Dose) and TD_{50} (Toxic Dose) are equivalent, synonymous terms. They may be used interchangeably.

(4) The correct answer is "d." Prozac®/fluoxetine has a wide therapeutic index. It can still cause problems, but it is considered a relatively safe medication.

(5) The correct answer is "b." On the dose-response curve, with dosing amounts increasing from left to right, moving right decreases the potency as the dose amount increases. Once again, everything in pharmacology is backward. If the dose increases, usually the potency decreases.

BONUS QUESTION ANSWER

The correct answer is "d," ED_{50}. The other choices are "plausible distractors" — not real terms. Potency is not as important; it is the strength of the drug. To compare the medications for "which drug works best to relieve worries," the effective dose is the winner.

Home, Sweet Home

Neurons and Neurotransmitters

Where do neurotransmitters live? Where do they come from? Where do they hang out? What is their address? At the risk of silliness, and for the sake of mental visualization, they live in the "city" and surrounding environs of a town called "Neuronville."

It is a populated city state with strict rules, all designed to protect the functioning unit's integrity: No political parties, competitive sports, or corporations. No gossip. All inhabitants mind their own business, hardly worry what other entities (organelles) are doing; yet one unit's actions facilitate or impede another and vice versa. Interestingly, the "people" of Neuronville only know what they do and are not concerned with other units' functions. It is all interdependent. One failure challenges the entire city design. It is all about working cooperatively and serving the ultimate good of the township of "Neuronville."

Given proper nourishment, Neuronville is self-sufficient — even has its own power plant, producing energy (mitochondrion). There is even a post office (Golgi bodies). Many roads, called dendrites, are all leading into the city center, and one big avenue heading away, out of town, the axon. There are train tracks (microtubules) moving essential materials all over the city. Within the city limits is the "cell body," a happening town square called the "nucleus" - which for convenience is always located in the center of the cell body. There are workers all around, maintaining the structural integrity of Neuronville.

Neuronville is an information-delivery hub, a messaging center, like a giant distribution center. Neuronville has one primary function: Transfer, release, and signal information up and down the body - at the proper speed and proper electrical strength. Messengers are called "chemical neurotransmitters," protected as if each one was a piece of certified mail, carefully packaged, and held in a vault-like collection container (vesicles) at the end of the axon in a special port called the "terminal button." With sufficient incentive (electrical impulses) moving down the tracks, the cargo containers (vesicles), packaged in the wrapping complex (Golgi bodies), release their chemical cargo (neurotransmitters) into a salty, fluid-filled synaptic sea (extracellular fluid) at the end of the main axonic thoroughfare.

Crossing this salty, synaptic sea (1/10,000th of an inch or 500 Å, angstroms wide), messages reach the appropriate recipient (receptor), and the process begins again at a new dendrite leading to another Neuronville. All total, there are approximately 100 billion (give or take a few billion) Neuronvilles duplicated throughout each human body, working steadily, just hanging out, doing what they are programmed to do.

There are foreigners, xenobiotics, trying to make Neuronville their adopted home. Drugs. Acting like newly manufactured neurotransmitters, if they somehow clear customs, passport control, and make it over the cytoplasmic wall (pharmacokinetics), they become a pharmacodynamic force all over the town of Neuronville. They have influence, can manipulate the system, modify in a single bound, mimic the good, block all things neuronally normal, and even maim or kill. All towns have heroes. All towns have villains. Neuronville is no exception.

Where do neurotransmitters hang out? Where do they originate? Until they are released, you will find neurotransmitters safely ensconced inside vesicles (from Latin, "vesicularis" meaning "little blister" or "little bladder"), or holding reservoir, safely protected in the terminal button at the end of the axon terminal, all waiting for just the right moment to be released into the salty synaptic space or cleft separating one township from another.

In the "real body," there are billions of neurons (from Greek, "neura," pertaining to the nervous system), and all neurons have three distinct features: (1) the cell body (soma), (2) dendrites, which look like trees (hence from Greek, "dendrites," meaning "pertaining to a tree") and (3) one axon (from Greek, "axon," meaning "axis"). Dendrites receive information; the axon conducts an electrical signal from the base of the soma (the axon hillock) all the way to the terminal button where the signal, if strong enough, will precipitate neurotransmitter release. It is organized as follows:

Dendrite → Soma → Axon → Cleft → Dendrite → Soma → Axon → REPEAT

Neurons are "sending and receiving machines." That is their purpose and what they do best. Before any neurotransmitter release, there must be a signal which causes neurotransmitter release from vesicles inside the terminal button. This initial signal is electrical, and if strong enough, neurotransmitters are released and flood the cleft or synapse. The synapse or space between neurons (they do not touch) is the area where nerve impulses are transmitted by one neuron and received by another's dendrites. Vesicles store millions of chemical molecules and release them when told. They do not falter. If the electrical signal is strong enough, vesicles secrete their protected neurotransmitters into the cleft. This beginning point is referred to as "presynaptic."

Once the released neurotransmitters enter the cleft, the sending portion terminates, and the receiving system begins. Neurotransmitters or molecules, released from the vesicles, make their way to receptor sites on the surface of the neuron. These receptor sites lie across the synaptic sea or cleft and are found on an adjacent neuron, referred to as postsynaptic.

Presynaptic (Sending) → Synaptic → Postsynaptic (Binding/Receiving)

To elaborate, presynaptic (sending, electrical) → synaptic (chemical neurotransmitter release into the open space) → postsynaptic (chemical neurons cross the cleft and bind to a receptor), then the process begins again.

In the final stage, neurotransmitters and receptors "hook up." The round peg finds the round receptor hole. The key finds the lock. The neurotransmitter takes a mate. The message is delivered. This hookup response can be agonizing or antagonizing. With affinity it can either stimulate or block. It can be excitatory; it can be inhibitory; or in the case of nicotine, both.

When a neuron is released from its presynaptic vesicle into the cleft, it is free; and four methods reflect mechanisms by which neurotransmitter action is terminated. Four things can happen. Neurotransmitters can (1) bind to a receptor and then dissociate or release; (2) diffuse; (3) transport/return to its presynaptic home via reuptake; or (4) be broken down or metabolized. They either do their job, wander off, break apart, or go home.

Bind to a receptor. Presynaptic vesicles "prime and dock" (move around the terminal button toward a point of exiting by a mechanism that is not clearly understood) and release their neurotransmitter cargo into the cleft or synapse. When the neuron reaches action potential (also known as depolarizing or "firing"), calcium rushes in, and neurotransmitters flood the cleft. The electrical process is completed, and neurotransmitters take on their full-fledged chemical responsibility. A neurotransmitter's purpose is to find a receptor, be the key in the lock, and activate that postsynaptic receptor. The result: Receptor is agonized (or antagonized), and conformational changes occur. Now the receptor, via intrinsic activity, responds as it is wired to do. The circle of neuronal life begins again.

Diffuse. This process affects all molecules in the body, not just neurotransmitters. Simply, the neurotransmitter "wanders off" or "drifts away" from the synaptic cleft; it can no longer activate a receptor and eventually makes its way to the kidneys (or it could go to the liver) where it is washed out of the system. This is another way that the body clears neurotransmitters from the cleft and resets. If it is a lipid-soluble neurotransmitter like nitric oxide (which does not play by any of the rules) or the endocannabinoids (similar to cannabis neurotransmitters), diffusion is the only mechanism by which they are removed from the cleft (Advokat, et al., 2019).

Transport or Reuptake. Neurotransmitters are recycled. This is the preferred method for "classical" (dopamine, norepinephrine, serotonin, histamine, glutamate and GABA) small-molecule neurotransmitters. EXCEPTION: ACETYLCHOLINE. It does not have a reuptake mechanism. Transport is a reabsorption method into the terminal button where the unused neurotransmitters are repackaged in a new vesicle, sequestered until the vesicle primes and docks, and then released again into the cleft – maybe this time it will find a receptor.

To elaborate, the MOA of Prozac®/fluoxetine is SSRI. Any drug whose mechanism of action ends in "RI" is a reuptake inhibitor drug. In this case, the drugs in this class block reuptake – "RI." Whether it is a "SSRI," "SNRI," "SARI," or "SNDRI," two key letters, "RI," reuptake or transport, explain that autoreceptors will be blocked with no egress from the cleft back into the terminal button.

Broken Down or Metabolized. Drugs are biotransformed by the liver (using Phase I (CYP450) or Phase II metabolism). Enzymes like MAO or COMT also break them down at the presynaptic, postsynaptic, and terminal button level.

Just like soup needs a spoon, a train needs a track, a prescriber needs a client, a neuron must have a neurotransmitter.

QUIZ TIME

(1) Most, not all, neurotransmitters tend to be released _____ into the _____.

    (a) postsynaptic; synapse

    (b) synaptically; presynaptic membrane

    (c) synaptically; postsynapse

    d. presynaptic; synapse

(2) What can happen when too many neurotransmitters are released into the cleft?

    (a) Party

    (b) They become super-agonists

    (c) Death

    (d) All are broken down

(3) If a patient has overdosed from an agonist, the medical treatment is to administer a(n) _____.

    (a) inverse agonist

    (b) antagonist

    (c) partial agonist

    (d) indirect agonist

(4) Which mechanism of neurotransmitter termination is like recycling?

    (a) reuptake

    (b) enzymatic degradation

    (c) diffusion

    (d) concentration gradient

(5) Action potential travelling down the axon, neurotransmitters released into the cleft, and then binding to the postsynaptic receptor is _____.

    (a) chemical, electrical, chemical

    (b) electrical, chemical, electrical

    (c) only chemical

    (d) only electrical

ANSWERS

(1) The correct answer is "d." Most neurotransmitters are released presynaptically into the synapse where they make their way to a postsynaptic receptor. The key word here is "most." There are neurotransmitters, like adenosine and the body's endogenous cannabinoids (endocannabinoids), which are released postsynaptic to presynaptic.

(2) The correct answer is "c." When too many neurotransmitters are released into the cleft, beyond the body's capacity to break them down quickly, they flood, saturate the receptors. The result can be organ spasm, shutdown, and death.

(3) The correct answer is "b." If a patient overdoses on a medication that is an agonist, then a receptor blocker is required to prevent the agonist from binding. Conversely, if a patient overdoses on an antagonist, then the treatment of choice is to prescribe a drug that floods the cleft with an agonist.

(4) The correct answer is "a." Reuptake and transport are synonymous terms. In each case, active neurotransmitters are deactivated when they move from the cleft, transported back into the terminal button via the reuptake apparatus and then repackaged in the vesicle with Vitamin C for freshness.

(5) The correct answer is "b." The axonal avenue operates electrically as current begins at the axon hillock, makes its way down the axon via saltatory (jumping from Node of Ranvier to Node of Ranvier) conduction. This "jumping" conduction is faster than the current moving steadily. At the terminal button, when an action potential is detected, the neuron depolarizes, and vesicle packages full of neurotransmitters are released into the cleft. Neurotransmitters are chemical. When neurotransmitters move across the cleft and bind to a postsynaptic receptor, then and only then does the electrical process begin anew. So, it is electrical, to chemical, to electrical. Or as the early 1900 pioneering researchers put it, "It is a spark to soup to spark."

Module 18

Back to the Future

Time for a Review

😊 😊 😊 😊 😊

Use it or lose it. This adage applies to psychopharmacology. Many a student passes the course with a cursory, passable 70%-75% understanding; but as routine kicks in, work in the clinic builds up, and daily life happens, there is a rapid rate of memory decay with this "so-much-to-remember" field. To that point, it is acknowledged that average prescribers have about 10 to 15 drugs they use routinely; good prescribers have about 20-25 pharmaceuticals in their quiver; and advanced practitioners are well versed in about 35 medications. However, in psychopharmacology alone, there are approximately 140+ medications available, making the concept of the right drug for the right patient even more difficult. So, time for a review – especially before launching into modules about "The Big Seven" neurotransmitters and the autonomic nervous system.

Psychopharmacology is a complicated field of study, but it is also like learning a new language. To do it well takes work, regular review, and a desire to keep up and stay current in the field – which is the most challenging aspiration of all.

First off, many people look at drug names, and trepidation kicks in. Remember, pronounce every syllable equally. Of course, mastering one name, does not mean that is the only name for that drug. Each drug has at least four names. The chemical, brand (always capitalized), and generic names do not change. Keeping up with the street names is a trick. Currently there are approximately 75 street names for cocaine and 150 everchanging, every-day-adding new descriptors for cannabis/marijuana. If one refers to any of these psychopharmacological agents as psychotropics, one will always be correct.

It is incredibly important to remember what pharmacologists call Rule 1: All drugs have a mechanism of action (MOA). For instance, Prozac®/fluoxetine's mechanism of action is SRI (Serotonin Reuptake Inhibitor). Once upon a time, it was called a SSRI (Specific SRI), but that is not necessary because there is nothing "specific" about this class of medications. Those in the know refer to them as SRIs. The problem here is that the SRI tells why you are prescribing the medication, but it does not advise complete drug action.

Two big opposite systems working in tandem, but against each other, are pharmacokinetics (Absorption-Distribution-Metabolism-Elimination) and pharmacodynamics (drug-receptor action). When the drug enters the system, pharmacokinetics kicks in with one purpose: To inactivate a xenobiotic and to biotransform it into something that cannot harm the system. If the drug makes it past the liver (bioavailability), the drug is now in control.

The most crucial point to remember is that the liver metabolizes (or breaks down), and the kidneys clear (first-order and zero-order kinetics). As a result, the liver and kidneys are the two most important organs when it comes to medications, and their health must always be considered before a medication is prescribed.

The family of Cytochrome P450 Enzymes are incredibly important to understand. Presently, there is a lot of marketing hype surrounding genetic tests that assess their presence or absence. While these tests are in their growth years, they represent a great hope for the future and should be considered as a means of protecting the patient from unwanted, toxic side effects.

What do CYP enzymes do? They break down substrates – in this case medications. With the proper amount of enzyme, drug doses enter the system with hoped-for, minimal problems. Suppose the patient is missing any of these CYP genomic enzymes, and the drug prescribed is a substrate (victim) of these enzymes. In that case, it is not broken down properly, and the patient can experience toxic side effects quickly. Additionally, some drugs are inducers and inhibitors of CYP enzymes. This means, if a drug is mixed with another drug that is either an inducer or an inhibitor, then the amount of the first medication will either be reduced or increased. Inducers can decrease the dose of another drug; inhibitors can increase the dose of another drug.

Prodrugs are special creatures. They are inactive drugs taken enterally (by mouth), and they require a healthy liver to biotransform and turn them into an active metabolite. Codeine is considered the prototypic prodrug. When it is taken, it is inactive, and then the liver turns it into 10% morphine. The point to remember: If the patient has a healthy liver. No problem. If the patient does not have a healthy liver, avoid prodrugs.

Ligands bind to receptors. The main ligands for psychopharmacology are "The Big Seven" neurotransmitters (Dopamine, DA; Norepinephrine, NE; Epinephrine, Epi; Serotonin, 5-HT; Acetylcholine, ACh; Gamma Aminobutyric Acid, GABA; and Glutamate, GLU). The two big receptor types are ionotropic and metabotropic (G coupled, 2nd messenger, a cascade of action) receptors. A neurotransmitter can agonize, antagonize, partially agonize, indirectly agonize, inversely agonize, and in some cases, both agonize and antagonize. To bind there must be a friendship, relationship, or an affinity between ligand and receptor. When the round peg finds the round hole, a conformational change happens, and the receptor responds with prewired behavior, intrinsic activity. When an antagonistic ligand blocks a receptor, the receptor stays in its resting state, but no agonist can bind while the receptor is plugged, so no intrinsic activity manifests.

The body has billions of best friends called neurons. Presynaptically, it is electrical; in the synaptic cleft, where neurotransmitters are released, it is chemical; when neurotransmitters find a postsynaptic receptor, the electrical system starts anew. When a neuron reaches action potential (electrical), neurotransmitters (chemicals) are released from their vesicles inside the terminal button into the synaptic cleft. Once released, one of four things can happen. The neurotransmitter can (1) bind to a receptor, (2) diffuse or wander off, (3) return to the terminal button from whence it came to be repackaged into a new vesicle, or (4) be metabolized by enzymes like MAO and COMT. The body insists that a neurotransmitter does its job; when it does, it is inactivated. If not, it is transported back into the terminal button or inactivated by enzymatic action. If too many neurotransmitters are released, harm can come to the body. In that case, the body's fail-safe mechanisms will kick in to protect itself.

Pharmacology is a study of negative feedback loops (like the thermostat in a home). When the temperature goes up, the heater shuts down; when the temperature goes down, the heater turns on. This is another example of pharmacological opposites like pharmacokinetics vs pharmacodynamics, agonists vs antagonists (inverse agonists), inducers vs inhibitors, association vs dissociation (drug binding), and perhaps the biggest opposite of all, the parasympathetic vs the sympathetic system.

To understand how one's nervous system functions, a thorough understanding of acetylcholine is required. Because acetylcholine, norepinephrine, and epinephrine are integral to the parasympathetic, sympathetic, and somatic systems, respectively, these nervous systems will be interspersed between "The Big Seven" neurotransmitter modules.

QUIZ TIME

(1) Of the four terms (psychoaffective, psychotherapeutic, psychoactive, and psychotropic) all psychoactives are psychotropics, but not all psychotropics are _____.

(2) Pharmacokinetics and pharmacodynamics explain what the _____ does to the drug and what the _____ does to the body, respectively.

   (a) body; drug

   (b) drug; body

   (c) drug; brain

   (d) body; brain

(3) Assuming first-order kinetics, if a drug is taken as prescribed, after _____ doses, a therapeutic, steady state is achieved.

   (a) 4.5

   (b) 5.5

   (c) 6

   (d) 6.5

(4) The liver _____, and the kidneys _____.

   (a) metabolizes/biotransform

   (b) metabolizes/clear

   (c) clears/metabolize

   (d) clears/bioavail

(5) Imagine a patient taking a form of birth control that is a substrate of a CYP450 enzyme. The patient mixes an over-the-counter medication, a CYP450 inducer of the same enzyme with their contraceptive. What will happen to the effectiveness of the contraceptive?

   (a) nothing

   (b) less contraception will find its way into circulation

   (c) more contraception will find its way into circulation

   (d) the over-the-counter medication will make the contraceptive more hydrophilic.

ANSWERS

(1) Psychoactive.

(2) The correct answer is "a." Pharmacokinetics is what the body does to the drug; and pharmacodynamics is what the drug does to the body.

(3) The correct answer is "b," 5.5. Remember, assuming first-order kinetics, once a medication is no longer needed, after 5.5 half-lives have passed, the drug is out of the system.

(4) The correct answer is "b." The liver metabolizes, and the kidneys clear.

(5) The correct answer is "b." Adding an inducer will metabolize more birth control medication. Less contraceptives will find their way into circulation, and pregnancy is conceivable.

All Hail the King

Acetylcholine

☺ ☺

Of "The Big Seven," serotonin is the most popular. Dopamine is the most desired. Histamine is the least thought about and not even included in the seven – but it should be. Norepinephrine and epinephrine are the most energizing. GABA is the most laid back, and glutamate is considered the master switch of the central nervous system. However, there is no doubt about it, acetylcholine is the most necessary and the "King of Neurotransmitters." Neither an eyelid nor little finger will move, nor any memory stored without acetylcholine. Without acetylcholine, one's heart would race out of control, and one would not be. It is that important.

Acetylcholine shares two standard features with other neurotransmitters. First, neurotransmitters do more than one thing. To explain, serotonin is integral to mood, but it is also important for appetite, apathy, pain, suicide, sleep, osteoblasts in the bone, and intraocular pressure. Even though this text focuses on neurotransmitters' action in relation to psychopharmacology, neurotransmitters work all over the body and not just in the brain.

Secondly, there is a balancing act, homeostasis among "The Big Seven" neurotransmitters. It is impossible to work with one and not "accidentally" manipulate the other six. Over one hundred neurotransmitters have been discovered;

no doubt there are homeostatic relationships between some or possibly all neurotransmitters. So, given the state of modern psychopharmacology, as mental health professionals and prescribers, "what we think we are doing" and "what we are actually doing" may be two different things. For instance, when serotonin is increased, dopamine is decreased. When dopamine is decreased, acetylcholine is increased. Continuing to reduce dopamine release will also decrease interneuronal glutamate and increase GABA. Prescribers must remember that when one gives a little pill to boost serotonin, the other six neurotransmitters are also being manipulated, indirectly agonized, or antagonized.

Acetylcholine was the first neurotransmitter to be identified in the peripheral nervous system and later in brain tissue. Henry Hallett Dale discovered naturally occurring acetylcholine in 1913, but Otto Loewi found its function. Loewi's discovery is the story of lore and a late night-dreaming, Easter weekend,1921 (McCoy & Tan, 2014). He had a dream, forgot the dream, re-dreamed the dream, remembered the dream, and followed his dream's instructions. Through early morning experimentation with frogs' hearts, he figured out the function of "vagusstoff," the original name for acetylcholine. Both scientists shared the 1936 Nobel Prize (McCoy & Tan, 2014; Parsons & Gannelin, 2006; Passani, et al., 2014; Tansey, 2006).

Recall that acetylcholine is a unique neurotransmitter in a class by itself. It is distinctive in that it is not a monoamine or amino acid. It is a non-amine-amine. Acetylcholine is essential for movement and memory (among other things). Movement does not happen without activation of acetylcholine; as a result, it is associated with voluntary muscle contractions. Additionally, and importantly for

124

psychopharmacologists, acetylcholine is required for memory consolidation and memory storage. While acetylcholine is distributed widely in the brain, the cell bodies of cholinergic neurons lie in the brain's basal forebrain (including the nucleus basalis of Meynert and substantia innominata), an area which significantly impacts learning and memory (Tansey, 2006).

There are two types of receptors associated with acetylcholine. They are called muscarinic receptors (the receptor is labelled with a capital "M") and nicotinic receptors (labelled with a capital "N"). They were originally named when early researchers were searching for natural substances, "chemicals," or anything that would agonize or stimulate these receptors. The muscarinic receptor was named because it was initially stimulated by a toxic alkaloid found in *Amanita muscaria* (Fly Agaric mushroom). Nicotinic receptors were named because they were responsive to tobacco's nicotine. Both muscarine and nicotine mimic acetylcholine ability to engage receptors. Acetylcholine is broken down rapidly (25,000 molecules a second) by acetylcholinesterase (AChE). The one-step breakdown and one-step creation (acetyl CoA + choline) of acetylcholine testifies to the importance of acetylcholine to the whole body.

If a patient is given a pill of acetylcholine, it will never leave the systemic circulation and move into organ systems. That is because acetylcholine is permanently charged (technically, it is a quaternary amine). It is polar and will be excreted in an unaltered form by the kidneys. To increase a person's level of acetylcholine, current drugs must block the breakdown of acetylcholine thereby indirectly increasing the level of acetylcholine throughout the entire system. This is how the memory/dementia medication, Aricept®/donepezil, works. The drug inhibits the breakdown of acetylcholine all over the body, including the small area in the brain thought to be related to memory enhancement (the basal forebrain).

What is the takeaway concerning acetylcholine? It is king. It is essential for movement and memory. Much will be said later about this non-amine-amine. Acetylcholine's importance in the parasympathetic, sympathetic, and somatic system cannot be understated. The modules devoted to these areas are often considered the foundation for pharmacology and follow a brief discussion of the monoamines and two amino acids, GABA and Glutamate.

QUIZ TIME

(1) Acetylcholine is a _____ charged, quaternary amine.

    (a) permanently

    (b) partially

    (c) hydroponically

    (d) temporarily

(2) Acetylcholine facilitates movement and _____.

    (a) appetite

    (b) apathy

    (c) sleep

    (d) memory

(3) Aricept®/donepezil's MOA is an _____.

    (a) acetylcholine agonist

    (b) acetylcholinesterase blocker

    (c) acetylcholine antagonist

    (d) acetylcholine inverse agonist

(4) True or False? Neurotransmitters do not have multiple functions. They only do one thing.

(5) Acetylcholine is a _____.

   (a) catecholamine

   (b) monoamine

   (c) indolamine

   (d) histamine

   (e) none of the above

ANSWERS

(1) The correct answer is "a." It is permanently charged. As a result, it stays inside systemic circulation, never enters organ systems until the kidneys clear it from the blood stream.

(2) The correct answer is "d." Two essential functions of acetylcholine are movement and learning/memory. Choices "a," "b," and "c" are associated with serotonin.

(3) The correct answer is "b." By blocking acetylcholinesterase (the enzyme that breaks down acetylcholine), it facilitates an increase of acetylcholine throughout the body. The result is not only an increase in acetylcholine in the brain's memory center but the entire body. Unwanted side effects are a result of this lack of specificity.

(4) The correct answer is "FALSE." A neurotransmitter working with various receptors on different organs and neural systems can elicit differing actions.

(5) The correct answer is "e." Acetylcholine is a permanently charged, quaternary amine.

Ticket to Ride

Acetylcholine and The Parasympathetic Nervous System (PNS)

☺

It is a 100% fact. Psychotropic medications affect the body's natural nervous system, which controls the body's internal state. The overall nervous system is composed of (1) the Central Nervous System and (2) The Peripheral Nervous System. The Peripheral Nervous System is divided into the Somatic and the Autonomic Nervous System (ANS). The Autonomic Nervous System (ANS) is further divided into the Parasympathetic Nervous System (PNS) and the Sympathetic Nervous System (SNS). The neurotransmitters (not a complete list) involved in these systems are listed in the following chart. More neurotransmitters are associated directly or indirectly with these systems, but for psychopharmacological reasons, these are vital.

NERVOUS SYSTEM ORGANIZATION & Vital Neurotransmitters			
Central Nervous System	Peripheral Nervous System		
	Somatic	Parasympathetic	Sympathetic
Histamine	Acetylcholine	Acetylcholine	Acetylcholine & Norepinephrine

Histamine and The Central Nervous System (CNS). Although not one of "The Big Seven," it is important. In a word, it is "central" to the Central Nervous System and generates sensory and motor experiences for the entire body. All "The Big Seven" neurotransmitters are represented in this system; however, histamine is often overlooked, except when one gets a stuffy nose. It was isolated from ergot mold in 1910 by Henry Hallett Dale and Patrick Laidlaw. They also discovered histamine's function and biological actions (Parsons & Gannelin, 2006).

The brain requires a steady stream of histamine, and the level of histamine determines consciousness. Agonizing histamine increases alertness; antagonizing it (an antihistamine) comes with drowsiness and a warning "not to use heavy machinery." Histamine may also play a part in addictive behaviors and neurological disorders like Parkinson's. In addition to acetylcholine, histamine likely plays a role in the maintenance of memories. Appetite is also affected by histamine. Remember that many psychotropic medications block histamine. The result: Consciousness is blocked, sleep is induced, and hunger increases. Histamine does not deserve its "step-child" status. It is an important neurotransmitter that must be considered when one prescribes psychopharmacological medications (Passani, et al., 2014).

Acetylcholine and The Parasympathetic System (PNS). It is de rigueur, more than essential, to understand the "play-by-the-rules" strict relationship between the Parasympathetic Nervous System (PNS) and acetylcholine. To demonstrate this importance, start with the name. Terms like "cholinergic" (shorthand for "acetyl-choline," an agonist) and "anticholinergic" (an antagonist) are used regularly. Agonizing or antagonizing, stimulating or blocking, the parasympathetic nervous system garners an amazingly helpful and/or detrimental set of actions and side effects. This is essential knowledge, and a drug side effect system that is heard repeatedly is called "anticholinergic." If one does not know this word's definition, then an obligatory understanding of at least ten (some life-threatening) side effects are lost in a clinician's ability to protect a patient.

The PNS is your rest, digest, breed, feed, and "take-out-the-trash" system. It is hard for many to believe that it is an overall far more powerful system than the sympathetic "flight or fight system." These are opposites. Parasympathetic = resting; sympathetic = flight.

When one is calm and resting, the parasympathetic nervous system (PNS) is in control. If a heart rate is around 70 beats per minute, thank the PNS. With no PNS in place, an average heartrate would be 100+ beats per minute. If the breathing rate is approximately twelve to sixteen breaths per minute, it is the PNS in control. If actively digesting your food, thank the PNS. Liquids flowing (e.g., saliva), and blood pressure on the mark? PNS rocks. Blood perfusion to visceral organs? PNS in control. Bowel movements regular? It is the PNS system that obliges. Able to see up close? Once again, thank the PNS. When one takes a cholinergic drug that stimulates, activates, agonizes, or indirectly agonizes the parasympathetic system, breathing decreases, heart rate decreases, blood pressure decreases, digestion and intestinal movements (e.g., diarrhea) increase.

When the PNS system is overly agonized or activated, here is what happens: "SLUDGEM" - Increased Salivation, Lacrimation (tearing), Urination, Defecation, GI distress, Emesis (the act of vomiting), and Miosis (excessive constriction of the pupil). In other words, everything that is resting, rests more. Every orifice leaks more. Every intestinal function moves faster and removes more waste.

Patients who take the drug Aricept®/donepezil for memory issues/dementia complain of "SLUDGEM." This drug's MOA is an AChEI – acetylcholinesterase inhibitor. It indirectly agonizes the parasympathetic system, blocks the breakdown of acetylcholine, and causes an increase of acetylcholine all over the body. Since the parasympathetic nervous system is activated by acetylcholine, any increase in acetylcholine will cause the resting state to "rest more," and the digestive state will become more active. As a result, approximately 1 in 12 patients stop this medication because of intolerable SLUDGEM.

If agonizing the parasympathetic nervous system creates SLUDGEM, then blocking the parasympathetic nervous system creates the opposite. Instead of salivation, dry mouth; teary eyes become dry eyes; urination is now urinary retention; defecation turns to constipation; GI distress turns into GI calm; emesis, now the probability of vomiting decreases. Agonizing the PNS yields miosis; antagonizing the PNS prompts mydriasis.

Agonizing the Parasympathetic Nervous System = SLUDGEM

Antagonizing the Parasympathetic Nervous System = DUCTS

When the parasympathetic system is blocked, when a person takes a drug that is an anticholinergic (cannot overemphasize the importance of understanding this term), the results are the opposite of SLUDGEM, turning it to DUCTS: Dry mouth,

Urinary retention, Constipation, Tachycardia, and Sedation (via central nervous system).

Acetylcholine activates the parasympathetic system. To understand the essential relationship between acetylcholine and its partner, the parasympathetic system, imagine traveling by train from Point A to Point B, but to get there, a change of neuronal trains is required. First, buy two tickets of acetylcholine. The first ticket (acetylcholine) is needed to get from the brain or spinal cord to the ganglia; the second ticket (also acetylcholine) will be used from the ganglia (train station or hub) to the effector organ. Now, board the train.

Riding the PNS railroad, to get from the brain (Point A) to the effector organ (Point B), the train stops at the ganglia (train station or hub). At the ganglia, switch trains. Acetylcholine functions as the token to ride both trains. Interestingly, it is a long neuronal ride from the brain to the ganglia but a short neuronal hop to the effector organ. It is a longer neuron from the brain to the ganglia and a shorter neuron from the ganglia to the effector organ. Two specialized terms are used to describe this changing trains neuronal handoff: Preganglia and postganglia.

Preganglia refers to neurons that come before the ganglia (small bundles of nerves originating in the brainstem or spinal cord), and postganglia refers to the neurons found after the ganglia. Put another way, preganglia neurons synapse with postganglionic neurons then go to the effector organ. First train = preganglia neurons; train station = ganglia; second train to destination = postganglia. Acetylcholine is the ticket to ride both.

Preganglia, acetylcholine is the neurotransmitter into a nicotinic receptor; postganglia, acetylcholine is the neurotransmitter that activates a muscarinic receptor. Because acetylcholine activates a muscarinic receptor, the parasympathetic system is also called the "cholinergic system" and the "muscarinic system."

It is a simple takeaway. Acetylcholine is essential for basic life functions. It is the king of the parasympathetic nervous system, the prime mover of the somatic nervous system (next module), and the sympathetic nervous system (even though norepinephrine is a close rival). Acetylcholine is essential for life functions. One cannot live without it.

PARASYMPATHETIC NERVOUS SYSTEM PRE- AND POSTGANGLIA

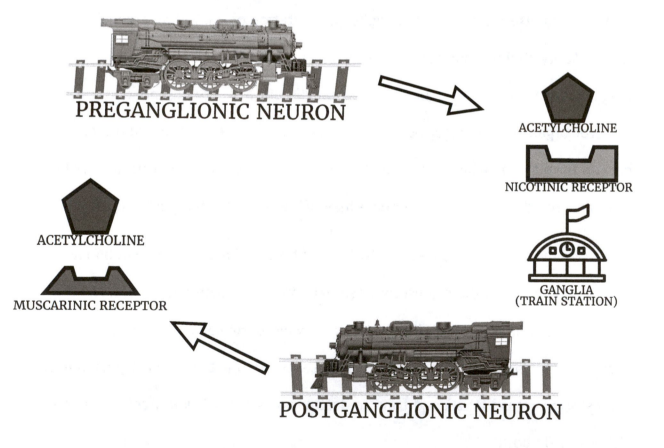

QUIZ TIME

(1) Grandpa is complaining about the side effects from his Aricept®/donepezil. Lots of urination, lots of defecation, and vision difficulties. Grandpa is suffering from

    (a) FUDGEM

    (b) SLUDGEM

    (c) DUCTS

    (d) OLDAGEM

(2) Another name for the parasympathetic system is _____.

    (a) muscarinic system

    (b) cholinergic system

    (c) adrenergic system

    (d) "a" and "c"

    (e) "a" and "b"

(3) _____ is the preganglia neurotransmitter into a nicotinic receptor in the PNS.

    (a) acetylcholine

    (b) norepinephrine

    (c) serotonin

    (d) histamine

(4) _____ is the postganglia neurotransmitter into a muscarinic receptor in the PNS.

    (a) acetylcholine

    (b) norepinephrine

    (c) serotonin

    (d) histamine

(5) Acetylcholine is

    (a) The King

    (b) The King

    (c) The King

    (d) The King

ANSWERS

(1) The correct answer is "b," SLUDGEM. The medication increases acetylcholine over the entire system. When the parasympathetic system is overly agonized, the side effects include salivation, lacrimation, urination, defecation, GI distress, emesis, and miosis. And for some, help with their memory issues.

(2) The correct answer is "e." The parasympathetic nervous system is named for acetylcholine (short-handed to "cholinergic") and the mushroom-like *amanita muscaria* (short-handed to "muscarinic"). These are the red and white polka-dotted mushrooms made famous in Disney's *Fantasia* (1940). In this movie, it is very clear why the hippos were dancing.

(3) The correct answer is "a." Acetylcholine is the preganglia neurotransmitter into a nicotinic, ionotropic receptor.

(4) The correct answer is "a." Acetylcholine is the postganglia neurotransmitter into a muscarinic, G protein, 2nd messenger receptor.

(5) The correct answer is "a," "b," "c," and "d." The King of Neurotransmitters, and never let it be forgotten. Long live the King!

The Prime Mover

Acetylcholine and The Somatic System

☺

Time to take the express train. Once again, acetylcholine is the ticket to ride, and this time the train goes directly from Point A to Point B with no change of trains. In other words, no ganglia. This train is high speed – much faster than the parasympathetic nervous system.

Like the parasympathetic and sympathetic nervous systems, the somatic system is the third part of the peripheral nervous system. It is associated with voluntary movement. Somatic system neurons project from the Central Nervous System (brain and spinal cord) and connect directly to muscles. Signals from muscles and sensory organs are relayed back and forth from brain to body to brain.

Imagine visiting the Acropolis in Athens. The perfect day to tour the Parthenon and visiting Greece seems appropriate since many pharmacological and neurological terms derive from Greek. For example, the word "somatic" comes from the Greek word, "somatikos," meaning "body." Being a typical tourist, strolling among the terrain, walking on ancient pavers, one is able to maintain body balance because the brain, spinal cord, cranial nerves, muscles, along with the somatic system are communicating and doing their job. The result: Upstanding, sauntering, and enjoying the visit. But wait, trouble is afoot.

While walking toward the gift shop, suddenly and unexpectedly, a not-so-level ancient paver "causes" an ankle to twist. It happens fast. Falling, tumbling headfirst, then suddenly regaining balance, responding thankfully for the body's ability to recover, and then walking away pretending that no one saw the somatic system two-step, quickstep, loss of balance, an almost fall.

What happened? The body's somatic system prevented a fall. The sensory system as a whole sensed an imminent fall. From visual system to inner ear, to every available sense, all sensory systems went into overdrive sending signals to the brain which immediately launched an all-out attack designed to do one thing: Help the somatic system move, stand erect, and maintain balance. The brain engaged muscles to act, allowing the body to twist, to turn, and to exercise a quick two-step dance designed to regain solid footing. The somatic system, working with multiple sensory and motor neurons, responded with a pronounced flow of acetylcholine, preventing a dangerous fall on ancient antiquities - not to mention hurting one's entire "somatikos." It is fast because there is no ganglia.

The somatic system includes spinal and cranial nerves are working together to provide sufficient sensory information to keep a body moving. It controls all voluntary muscles and reflex arcs (e.g., patellar reflex — tap your knee and leg jumps; touch your hand to a hot plate, and the hand pulls itself away quickly).

Time for a somatic system train ride. You are leaving Point A heading to Point B. This is a voluntary decision, so the somatic railway is the transportation mode of choice. There is no ganglia, and the neural response will utilize the neurotransmitter acetylcholine which will interact not with a traditional receptor, but acetylcholine will be released at the neuromuscular junction. This makes it fast.

A neuromuscular junction (NMJ) is also called a myoneural junction (the prefix "myo" from Greek meaning "muscle"). In medical parlance, anytime the prefix "my" or "myo" is added to a word, it translates to "muscle." The NMJ is not a traditional receptor but a chemical synapse between a motor neuron and a muscle fiber. While the NMJ is analogous to a receptor, it is just not quite the same. It is amazingly fast because the motor neuron transmits a signal directly to the muscle fiber causing a contraction.

Many types of venom work on the somatic nervous system. Imagine a poor little, unsuspecting mouse ambling in the forest. Little does the baby mouse know that a venomous snake is quietly lying-in wait. The creature slithers and strikes the mouse, and in one bite delivers a lethal dose of venom. Many reptiles' poisons work by blocking acetylcholine's breakdown in the cleft — just like Grandpa's Aricept®/donepezil. A dramatic increase in acetylcholine at the somatic system's neuromuscular junction causes the mouse's muscles to contract. The mouse cannot run. The somatic system train is derailed. Additionally, this type of venom causes organ spasms and ultimately organ failures.

What is the moral of the story? If the mouse had seen the snake, its somatic system (and sympathetic nervous system) would have given him the ability and the energy to run away. Unlike the parasympathetic and sympathetic systems, there is no ganglia. This makes the somatic system fast, amazingly fast. Since the mouse did not see the snake, the venom caused an increase in acetylcholine in the somatic system, causing muscles to contract, paralyze, and be served up as snake food. The mouse will also become very juicy. An acetylcholine agonist creates SLUDGEM: Salivation, Lacrimation, Urination, Defecation, GI Distress, Emesis, and Miosis.

Back to grandpa, his memory problems, and his medication, Aricept®/donepezil. Grandpa has been complaining about salivation, lacrimation, urination, defecation, GI distress, emesis, and miosis, but now a new complaint has been added to his SLUDGEM. He is complaining of muscle cramps because his somatic muscle system is overly agonized with acetylcholine.

The mechanism of action of his memory medication, Aricept®/donepezil, is an acetylcholinesterase inhibitor. It blocks the enzyme which breaks down acetylcholine, so acetylcholine builds up all over the body. This includes the somatic system. The somatic system controls voluntary muscle movements. When it is overly agonized, the result is cramps for gramps.

Acetylcholine is the king of neurotransmitters. It is the primary activator in the somatic, parasympathetic, pre-ganglia sympathetic, and sympathetic innervation of the adrenal medulla. Without acetylcholine so much that makes us human would not be possible (Whalen, et al., 2019).

SOMATIC NERVOUS SYSTEM NEUROTRANSMITTER AND NEUROMUSCULAR JUNCTION

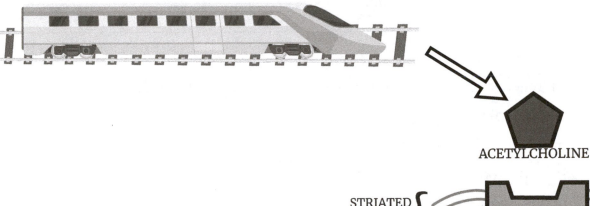

QUIZ TIME

(1) Which of the following systems does NOT require a ganglia.

     (a) sympathetic nervous system

     (b) parasympathetic nervous system

     (c) somatic nervous system

     (d) muscarinic system

(2) In the somatic system, _____ is the neurotransmitter released into the _____.

     (a) acetylcholine/neuromuscular junction

     (b) serotonin/neuromuscular junction

     (c) acetylcholine/muscarinic receptor

     (d) norepinephrine/metabotropic receptor

(3) Grandpa is continually complaining about the side effects from his Aricept®/donepezil. Lots of urination, lots of defecation, and difficulty seeing. Poor grandpa is suffering from _____.

     (a) FUDGEM

     (b) SLUDGEM

     (c) DUCTS

     (d) CRAMPEM

(4) Among other things, acetylcholine facilitates memory and _____.

     (a) appetite

     (b) apathy

     (c) sleep

     (d) movement

(5) Which mechanism of neurotransmitter termination is like recycling?

     (a) reuptake

     (b) enzymatic degradation

     (c) diffusion

     (d) concentration gradient

ANSWERS

(1) The correct answer is "c." The somatic system, unlike the parasympathetic and sympathetic system, does NOT have a ganglia. The result: Fast, very fast.

(2) The correct answer is "a." The somatic system's preferred neurotransmitter is acetylcholine, and it is released into the chemical synapse at the neuromuscular junction.

(3) The correct answer is "b." The medication increases acetylcholine over the entire system. When the parasympathetic system is overly agonized, the side effects include salivation, lacrimation, urination, defecation, GI distress, emesis, and miosis. And for some, help with their memory issues.

(4) The correct answer is "d." Acetylcholine innervates the somatic system, which is integral to facilitate movement, to stand tall, and to maintain balance.

(5) The correct answer is "a." Reuptake, transport, and recycle are equivalent terms. In each case, active neurotransmitters are deactivated when they are moved from the cleft, transported into the terminal button via the reuptake apparatus and then repackaged in the vesicle with Vitamin C, ascorbic acid, for freshness.

Module 22

The Power behind the Throne

Norepinephrine

☺ ☺

Norepinephrine and epinephrine are two remarkable neurotransmitters. They are the power source that causes a body to move fast, think fast, and get out of physical trouble. If acetylcholine is the king, then norepinephrine and epinephrine are the powers behind the throne.

Epinephrine was discovered in 1894 by Edward Albert Schafer and George Oliver. They showed the effects of adrenal gland extracts and found its impact on heart rate and blood pressure. In 1901, it was purified by Takamine Jokichi (Arthur, 2015; Rao, 2019). Compared to epinephrine, the discovery of norepinephrine is a modern discovery. Swedish biologist Ulf von Euler discovered norepinephrine in 1946 and won the Nobel Prize for his discovery. However, it was Marthe Louise Vogt, who in 1954, discovered norepinephrine in the brain. This was a vital discovery for psychopharmacology. It was not until 1964 that the specific seat of norepinephrine, the locus coeruleus, would be discovered within the human brain. Putting this together, the 1950s and 1960s were seminal years for neurology and pharmacology (Parascandola, 2010).

Norepinephrine is more technically referred to as noradrenaline. Additionally, epinephrine can also be referred to as adrenaline which is derived from Latin roots meaning "at or alongside the kidneys." In the United States, norepinephrine and epinephrine are used rather than noradrenaline and adrenaline, respectively. However, norepinephrine is the preferred international nonproprietary name. The body system that responds to these neurotransmitters is referred to as noradrenergic. Similarities aside, there are differences between norepinephrine and epinephrine.

First off, they are structurally related. When there are two similar words (epinephrine and norepinephrine), and one of them begins with "nor," they are similar, structural analogs with differing effects. Norepinephrine has more specific actions; epinephrine has more wide-ranging effects. Additionally, norepinephrine is continuously released into the body at low levels, whereas epinephrine is only released during fearful, frightening, stressful times.

Norepinephrine is the most common neurotransmitter in the sympathetic nervous system, made inside the nerve, stored in the vesicles, and released with action potential; however, epinephrine is released by the adrenal medulla and acts more like a hormone. However, the adrenal medulla can also make small amounts of norepinephrine, and epinephrine is made from norepinephrine. One thing that makes norepinephrine and epinephrine unique is that they are both hormones and neurotransmitters. Both are used in the flight-or-fight response to maximize the full strength and force of the sympathetic nervous system.

Norepinephrine/noradrenaline is integral in the function of the sympathetic nervous system. While the parasympathetic nervous system is your rest/digest/feed/breed/take-out-the-trash system, the sympathetic nervous system is responsible for the "Five F's" — Flight, Fight, Fright, Fornicate, and Freeze. Whatever the parasympathetic system controls, the sympathetic exerts the opposite (Roelofs, 2017; Schmidt, et al. 2008). To put it another way, the sympathetic system is antagonistic (the competitor) to the parasympathetic system. When the sympathetic system takes over and flight or fight kicks in, it is powerful; but the parasympathetic system is, in the long run, more powerful and regulatory than the sympathetic system. When necessary, it can even overpower the sympathetic nervous system.

Not a pleasant scenario but imagine a lurking danger. It is norepinephrine and epinephrine to the rescue. Primitive responses assume command, and the body will quickly mobilize (bullet train) to get out of harm's way. Sometimes it is so fast that it is not even processed consciously. That is the flight, fright, and if one must, fight reaction. Quicker than a wink, blood pressure goes up. Heart rate increases. Blood is shunted from the visceral organs to muscles providing more energy for the quick getaway. Vision changes so one can see far away. Up close vision (miosis) is not essential now. It is the distance from the threat that counts. Appetite goes away; eating is not necessary when running from a predator. The flight-fright-fight aspect is designed to do one thing – get out of trouble, flee the threat, and avoid potential harm. The sympathetic system will fight if it must, but its main task is to get away safely and fight another day.

Why an individual freezes when threatened is one of the great biological mysteries. Why, given imminent harm, is a person paralyzed with fear? Theories attempt to explain. Maybe the freeze response is more of an extension of a surprise response. When an unexpected event occurs, some will stop and evaluate the scene

145

to decide whether to run. Maybe paralyzing fear is equivalent to other animals "playing dead." Some posit this is a throwback to prehistoric times, when there is no way to escape, so play dead. Interestingly, the freeze response is unconscious; play dead, and hope the predator will lose interest and wander away. It may have a functional purpose, but it is bothersome as one recovers from the threat. It can be personally threatening in that the victim begins to experience self-doubt over their inability to handle the situation (Roelofs, 2017; Schmidt, et al., 2008).

Flight, fright, fight, freeze, what "F" is left? Fornication (from Latin, "fornix," meaning "vaulted chamber" or "arch" because those who received remuneration for sexual congress would wait for their customers out of the rain under vaulted ceilings in ancient Rome) as used here refers to sexual activity in general. No one system – parasympathetic or sympathetic — is responsible for the complete sexual response; they divide up responsibilities. The parasympathetic system is responsible for the male and clitoral erections; the sympathetic system is responsible for the male and female orgasmic response. These two systems in tandem create the energized sexual response. However, this distinction is vital because numerous psychopharmacological drugs have sexual difficulties as a significant side effect.

What is the takeaway? The sympathetic system is also called the adrenergic system because of its use of adrenaline and noradrenaline (epinephrine and norepinephrine, respectively). Its receptors (covered in the next module) are called "adrenergic receptors." Using the lock and key example, the key is norepinephrine; the lock will be any number of adrenergic receptors. If this system is agonized, a flight-fight response is engaged. If it is antagonized (or blocked), then a response like the parasympathetic system (rest and digest) will appear.

When the dominant resting system, the parasympathetic system, is activated, the sympathetic system is relegated to the background. When the sympathetic system is activated and foreground, the parasympathetic system is relegated to the background. The two systems CANNOT be active at the same time. If a medication agonizes the parasympathetic system, then the sympathetic system is pseudo-blocked. If the sympathetic system is agonized, then the parasympathetic system is pseudo-blocked. When this happens, side effects (DUCTS or SLUDGEM) kick in accordingly.

If a medication agonizes the parasympathetic system, the side effects are SLUDGEM (salivation, lacrimation, urination, defecation, GI distress, emesis, and miosis), and the sympathetic nervous system is pseudo-blocked.

Suppose the parasympathetic nervous system is blocked by an anticholinergic medication (a cholinergic or muscarinic system blocker). In that case, the side effect profile is DUCTS (dry mouth, urinary retention, constipation, tachycardia, and sedation). There are more side effects to this list that are not a part of the "DUCTS" mnemonic. These additional side effects include decreased sweating, confusion, visual problems, hallucinations, and diaphragmatic paralysis in a way that is not clearly understood.

Notice the inverse relationship between parasympathetic and sympathetic, SLUDGEM and DUCTS. If the sympathetic system is agonized and the parasympathetic system is pseudo-blocked, then the side effect profile is DUCTS. To explain, when one is in a threatening situation and the sympathetic system is activated, one experiences dry mouth, urinary retention, constipation, tachycardia (rapid heartbeat), and sedation. Inversely, if the sympathetic system is antagonized or blocked, the parasympathetic nervous system is pseudo-agonized, the physiological reactions include SLUDGEM – salivation, lacrimation, urination, defecation, GI distress, emesis, and miosis.

However, the question always comes up, why do some people "pee in their pants" when they are afraid? The answer, there are two detrusor muscles that form a bladder wall layer. Both maintain or hold urine. One is voluntary; the sympathetic nervous system controls the other. When activated, fear can prompt the sympathetic nervous system to jettison any unnecessary fluid weight.

Here is the "magic formula" that is incontrovertibly the foundation for ALL of pharmacology and psychopharmacology.

PNS Agonist→ SNS Pseudo-Antagonized/Blocked = SLUDGEM

SNS Agonist → PNS Pseudo-Antagonized/Blocked = DUCTS

NERVOUS SYSTEM/MEDICATION DRUG ACTIONS

DRUG ACTION	PARASYMPATHETIC (PNS)	SYMPATHETIC (SNS)
AGONIST/Stimulate or Indirect Agonist	SLUDGEM	DUCTS
ANTAGONIST/Block	DUCTS	SLUDGEM

Many psychopharmacological drugs affect the parasympathetic and sympathetic nervous systems as either agonists (typically indirect agonists like Aricept®/donepezil) or antagonists. As a result, numerous medications have a side effect profile which corresponds to SLUDGEM or DUCTS.

QUIZ TIME

(1) Epinephrine is _____, and norepinephrine is _____.

    (a) noradrenaline; adrenaline

    (b) norepinephrine; adrenaline

    (c) adrenaline; noradrenaline

    (d) an indolamine; a catecholamine

(2) Grandma's medication is an indirect agonist of the PNS. This means the drug pseudo-blocks the SNS. Grandma is continuously complaining of

    (a) SLUDGEM

    (b) DUCTS

    (c) heart palpitations

    (d) urinary retention

(3) DUCTS medication side effects include:

    (a) decreased sweating

    (b) confusion

    (c) visual problems

    (d) diaphragmatic paralysis

    (e) all the above

(4) The PNS and SNS are also called _____ and _____, respectively.

    (a) adrenergic; cholinergic

    (b) cholinergic; adrenergic

    (c) muscarinic; adrenergic

    (d) "b" and "c" are correct

(5) Circle the correct answer.

NERVOUS SYSTEM/MEDICATION ACTION		
DRUG ACTION	SYMPATHETIC (SNS)	PARASYMPATHETIC (PNS)
ANTAGONIZE	SLUDGEM or DUCTS?	SLUDGEM or DUCTS?
AGONIZE	SLUDGEM or DUCTS?	SLUDGEM or DUCTS?

ANSWERS

(1) The correct answer is "c." Epinephrine is synonymous for adrenaline; and norepinephrine is another term for noradrenaline.

(2) The correct answer is "a." Chances are grandma is taking Aricept®/donepezil which is an acetylcholinesterase inhibitor (AChEI), an indirect agonist of the parasympathetic nervous system. As a result, grandma is complaining of SLUDGEM – salivation, lacrimation, urination, defecation, gastrointestinal distress, emesis, and miosis. The odds are strong that grandma will discontinue this medication due to its side effect profile.

(3) The answer is "e," all the above. DUCTS is shorthand for Dry mouth, Urinary retention, Constipation, Tachycardia, increased Sedation. However, those in the know understand that anticholinergic side effects include DUCTS plus decreased sweating, confusion, visual problems, hallucinations, and diaphragmatic paralysis. So, it is "DUCTSCVHD" – that is why more people are familiar with DUCTS than the full cadre of side effects.

(4) The correct answer is "d." The PNS is also called the muscarinic and cholinergic systems, and the SNS is often referred to as the adrenergic system.

(5)

NERVOUS SYSTEM/MEDICATION DRUG ACTIONS		
DRUG ACTION	SYMPATHETIC (SNS)	PARASYMPATHETIC (PNS)
ANTAGONIZE	SLUDGEM	DUCTS
AGONIST	DUCTS	SLUDGEM

When the PNS is agonized, the result is SLUDGEM. When the SNS is agonized, the result is DUCTS. Look for tachycardia as a clue. When the SNS is activated, the only rapid heartbeat in DUCTS or SLUDGEM is the "T" in "tachycardia" (DUCTS). So, activating the SNS will also yield DUCTS. There is no rapid heartbeat when one agonizes the "resting" parasympathetic system. Instead, a slow heartbeat, bradycardia, is expected. Only when the parasympathetic system is blocked or pseudo-blocked, will tachycardia (the "T" in DUCTS) emerge.

Turbocharged

Epinephrine

😊😊

The early bird may get the worm, but it is the bird with the fastest, turbocharged sympathetic nervous system which not only gets the first worm, eats the worm, avoids predators, and lives to tweet about it. All courtesy of norepinephrine and epinephrine. That is how the sympathetic nervous system and its partner, the adrenal medulla, work for a creature's welfare. The sympathetic nervous system is fast, norepinephrine/noradrenaline fast. However, there is a quicker, faster, more powerful, turbocharged system. The king of neurotransmitters, acetylcholine, can command a mild-mannered adrenal gland (floating on top of the kidneys) to activate. Now fast-moving, turbocharged epinephrine (adrenaline) courses through the entire system. As a result, the "little engine that could" is now a turbocharged train. Putting these systems together, when the little engine is resting up at the train stations, it is fueled by the parasympathetic system. As it starts to move, the somatic system kicks in. To move faster and/or respond to a threat or get out of harm's way, that is the sympathetic system's responsibility. Finally, if there is a life threat to the "little train," epinephrine is the turbocharger that can turn it into an even faster train.

First, the sympathetic system, followed by sympathetic innervation of the adrenal medulla. Remember the ganglia? The sympathetic nervous system has a ganglia, too. To understand how it works, imagine getting from Point A to Point B fast (but not turbocharged train FAST). Two neurotransmitters will be involved, and a change of trains at the ganglia. Preganglia in the sympathetic nervous system, acetylcholine is released into a nicotinic receptor. Postganglia, norepinephrine is released into an adrenergic receptor on the effector organs. The result: ready to fight or flee.

However, sometimes even a fast system needs help in a life-threatening situation. Acetylcholine commands the adrenal medulla to release turbocharged epinephrine (adrenaline); this is called "sympathetic innervation of the adrenal medulla." This activation of the adrenal medulla happens fast, extremely fast – no thought required. Epinephrine secreted directly into the bloodstream turbocharges the sympathetic nervous system and is incredibly important for dealing with life-threatening, stressful situations.

It is important to always refer to this part of the adrenal gland as the "adrenal medulla." The adrenal glands, which float on top of the kidneys, are composed of a "medulla" and a "cortex." Each has a secretory function. It is the adrenal medulla that secretes epinephrine (80%) and norepinephrine (20%) into the system. However, this is not the only medulla (Latin for "pith" or "marrow," because like marrow, the adrenal medulla is in the center of the adrenal gland) in the body. The brain has a medulla oblongata. There is a renal medulla, the innermost part of the kidney. Hairs have a medulla, the innermost part of the hair shaft. As a result of numerous "medullas" in the body, it is important to always use both terms, like a first and last name, adrenal medulla, when describing this sympathetic innervating system. It is equally important to remember that the adrenal medulla secretes epinephrine and norepinephrine, and the adrenal cortex secretes the glucocorticoids. Confusing the two, adrenal medulla and adrenal cortex is a common mistake.

Here there are no train stations, no ganglia, no getting from Point A to Point B. When acetylcholine commands the innervation of the adrenal medulla, it is a "Do it and do it now!" command. Acetylcholine signals a nicotinic receptor on the adrenal medulla to release epinephrine into the blood. Because epinephrine is both a neurotransmitter and hormone, it circulates body wide activating adrenergic (derived from the word "adrenaline") receptors. That quick. That simple. That efficient.

Here is how it all works. Imagine a beautiful day, taking a leisurely walk in the park. The parasympathetic system, courtesy of acetylcholine, oversees basic day-to-day basal functioning – resting, digesting, appropriate heartbeat, blood pressure, all is right with the body world. An ambling, upright, relaxing walk is courtesy of the somatic system (and numerous cranial nerves). Suddenly, one hears footsteps approaching from behind. Perceiving a threat, the brain signals and steps quicken. Protective instincts prompt the sympathetic nervous system to gear up. Norepinephrine is released, just in case. The rest-and-digest system is now pseudo-blocked by the sympathetic system. The stranger is carrying a knife. The sympathetic system kicks in full force. Time for the turbocharged train.

SYMPATHETIC NERVOUS SYSTEM PRE- AND POSTGANGLIA

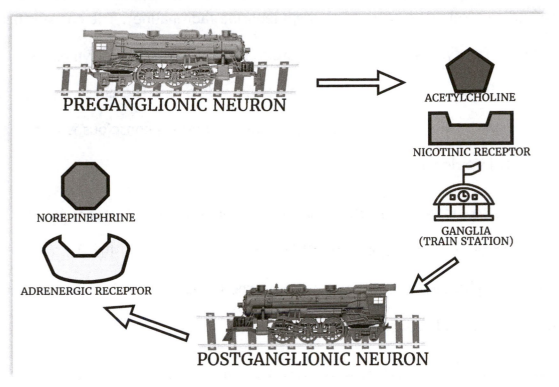

SYMPATHETIC INNERVATION OF THE ADRENAL MEDULLA

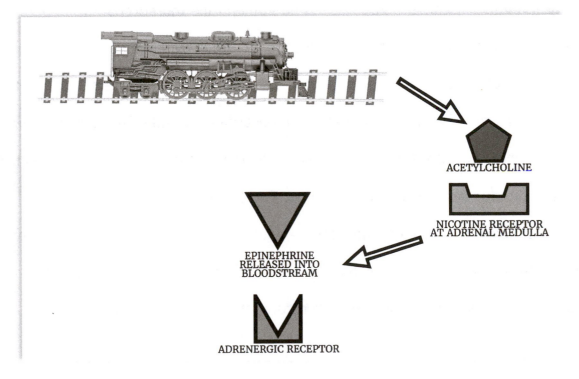

Acetylcholine activates nicotinic receptors on the adrenal medulla, and the adrenal medulla releases turbocharged epinephrine (adrenaline). The chase is on; run do not think. The sympathetic nervous system, somatic system, and sympathetic innervation of the adrenal medulla are operating wide open. The parasympathetic nervous system is relegated to the background. Now nothing conscious is going on; run and run fast. Life depends on it.

Thanks to turbocharged epinephrine, the would-be perpetrator is evaded. Now while hiding from the threat, still breathing hard, panting, out-of-breath, heart pounding, thoughts racing, it takes a while for the parasympathetic system to regain complete control and return the body to normal.

During the threat, a flood of neurotransmitters (epinephrine, norepinephrine, acetylcholine) was thrown into the body system — all designed to give energy and to activate a flight-or-fight response. Excess norepinephrine (noradrenaline) and epinephrine (adrenaline) are still coursing through the body's veins. A moment ago, they were needed; now they are excess, superfluous neurotransmitters. Heart rate and breathing are slowly returning to normal because MAO and COMT are breaking down excessive neurotransmitters (NE, DA, Epi) thrown into the system. It takes time, but the enzymatic system gets the job done. The system will not allow excess neurotransmitters to remain in the cleft stimulating receptors that do not need to be enabled. When these excess catecholamines are broken down or metabolized completely, the parasympathetic system returns the body to its resting state.

Finally, courtesy of enzymatic breakdown, the parasympathetic nervous system regains control, and normal homeostasis returns. However, the sympathetic nervous system patiently waits, just to serve and to protect all over again.

QUIZ TIME

(1) Epinephrine is secreted from the adrenal _____.

    (a) gland

    (b) cortex

    (c) medulla

    (d) shaft

(2) Which of the following systems does NOT have a ganglia?

    (a) parasympathetic nervous system

    (b) sympathetic nervous system

    (c) somatic system

    (d) none of the above

(3) Which neurotransmitter innervates the adrenal gland and triggers the release of epinephrine?

    (a) acetylcholine

    (b) histamine

    (c) serotonin

    (d) norepinephrine

(4) True or False? When the adrenal medulla is innervated, it releases equal amounts of norepinephrine and epinephrine.

(5) Match the following:

_____ Epinephrine	A. Glucocorticoids
_____ Parasympathetic	B. Muscle Contractions
_____ Somatic	C. Sympathetic Nervous System
_____ Norepinephrine	D. Feed
_____ Adrenal Cortex	E. Hormone

ANSWERS

(1) The correct answer is "c," adrenal medulla. Admittedly, "a" is a correct answer, but not as precise. It is important to remember that the "adrenal medulla" is the source of epinephrine.

(2) The correct answer is "c," somatic system.

(3) The correct answer is "a," The King. Acetylcholine is released into a nicotinic receptor at the adrenal medulla, and epinephrine is released into the blood where it interacts with adrenergic receptors throughout the body system.

(4) FALSE. When the adrenal medulla is innervated by acetylcholine, it releases 80% epinephrine and 20% norepinephrine.

(5) MATCHING

Epinephrine & E, Hormone. Recall that epinephrine is released directly into the bloodstream by the adrenal medulla, thus qualifying it as a hormone and a neurotransmitter. Additionally, norepinephrine is also released from the adrenal medulla and acts as a hormone, too.

Parasympathetic & D, Feed. The parasympathetic nervous system is the rest/digest/breed/feed/take-out-the-trash system.

Somatic & B, Muscle Contractions. When acetylcholine innervates a nicotinic receptor (N_m) at the neuromuscular junction, muscles contract.

Norepinephrine & C, Sympathetic Nervous System, postganglia norepinephrine into an adrenergic receptor elicits the "Five F's" — flight, fright, fight, fornicate, and freeze.

Adrenal Cortex & A, Glucocorticoids. The adrenal medulla and adrenal cortex are easily confused. The adrenal medulla releases epinephrine (80%) and norepinephrine (20%), and the adrenal cortex secretes the glucocorticoids.

The Big D

Dopamine

😊 😊 😊 😊 😊

It is the neurotransmitter people crave; and when they get it, it is never enough. It is the neurotransmitter responsible for love, and it is the neurotransmitter or lack thereof responsible for the pain when one breaks up. It is all things hoped for; once achieved, it rapidly becomes all things lost. It is the fuel of addiction, the thrill of the chase, the agony of failure, and the thirst for success. Often referenced by unwitting talk-show hosts, it is equated with the pursuit of happiness.

Chemically, it is a large, neutral, neither good-nor-bad amine; frankly, its actions are anything but neutral. Like Janus, the Roman god of duality, it is both angelic and sinister, which dopaminergic face is exhibited depends on a user's needs. Manufactured from dihydroxyphenylalanine, this miraculous neurotransmitter is dopamine. Or as psychopharmacologists affectionately call it: The "Big D."

Arvid Carlsson discovered the role of dopamine in 1957. Prior to the discovery of its function, it had been identified as an intermediate of epinephrine and norepinephrine synthesized from the precursor amino acid tyrosine. Prior to Carlsson's discovery, it was thought unimportant. Additionally, dopamine was not discovered in the brain until 1959 in the substantia nigra and ventral tegmental area (Bjorklund & Dunnett, 2007).

Dopamine is not just for the brain. Do not be fooled that dopamine is only for reward or mood. Remember the rule that all neurotransmitters do more than one thing? Dopamine proves this adage true. One example is dopamine's role as a treatment for cardiogenic shock (a rare condition, often fatal if not treated immediately, and mostly – but not always — caused by a severe heart attack). This is a situation in which the heart is suddenly unable to pump enough blood to meet the body's needs. To explain, recall that dopamine is part of the catecholamine family ("CATS live in the DEN"), and dopamine is a precursor (a substance that comes before and facilitates the creation of another substance) of norepinephrine and epinephrine. Administering dopamine during cardiogenic shock causes renal, mesenteric, and coronary beds to vasodilate, improving blood flow. (Note: The term "bed" is used to describe all blood vessels involved in an organ or region of the body). At higher doses dopamine can cause arterial vasoconstriction and an increase in blood pressure.

Note the following chart concerning other dopaminergic actions outside the central nervous system in the peripheral nervous system.

DOPAMINERGIC ACTIONS IN THE PERIPHERAL NERVOUS SYSTEM
↑ NA+ excretion in kidneys (Jiang, et al., 2016)
↓ Insulin in pancreas (Liu, et al., 2020)
↓ GI motility, protects intestinal mucosa (Li, et al., 2019; Li, et al., 2018)
↑ Urine output (Friedrich, et al., 2005)
↓ Activity of lymphocytes initiating immune response (Nasi, et al., 2019)

In the central nervous system, when most think of dopamine, three big behavioral and emotional aspects come to mind. First, dopamine is important in motor control – the ability to start, to stop, and to move gracefully, especially as it relates to fine motor movements. Deficiencies in motor control are clearly seen in diseases like Parkinson's, resulting from dramatic reduction of dopamine in the midbrain's substantia nigra (from Latin "black substance"). There is no doubt that dopamine is the neurotransmitter of choice for reward and smooth fine-motor movements.

Second, dopamine plays a role in emotion. It is thought (within the bounds of theory) that dopamine release, especially in the neocortex and limbic system, may be the root of schizophrenia. For instance, drugs designed to treat the psychoses, Thorazine®/chlorpromazine and Haldol®/haloperidol, block the number two dopamine receptors.

Finally, dopamine plays a significant role in reward and craving. Dopamine may be the motivating force behind any craving. Drugs that agonize dopamine specifically affect the nucleus accumbens septi (Latin for "nucleus adjacent to the septum"). Today, the brain's reward center is more simply referred to as the nucleus accumbens (NAc or NAcc). When it is agonized, all pleasure is released. It is the neurotransmitter that makes work seem effortless.

To that point, for years it was taught that dopamine is the neurotransmitter responsible for "The Five R's" – (1) Randy (sexual arousal), (2) Risky, (3) Rowdy, (4) Rewarding, and (5) Raunchy. No doubt excess dopamine makes an individual "frisky."

Too much dopamine leads to poor choices. It is rewarding and self-reinforcing. Under the influence of dopamine, one is likely to be more disinhibited, even vulgar. However, it is likely that these "Five R's" are more likely the means to a pleasurable end. Or to put it another way, the "Five R's" are behaviors or tactics that lead to getting what one wants. Dopamine pushes, "Go get it. No matter what one must do to get it, just do it." Because dopamine is so rewarding and powerful, one does not think beyond that point. Dopamine is the wanting beyond reason, and the "Five R's" are the "wanton" behaviors that manifest because of chemical cravings. This power is called "incentive salience."

Continuing the concept of "incentive salience", dopamine in the central nervous system is the neurotransmitter that makes work seem effortless. Additionally, increases in dopamine can affect the striatum's spiny neurons, causing the perception of time to pass quickly ((Meyer & Quenzer, 2019; Mitchell, et al., 2018; Marinho, et al., 2018).

To use a "randy," "frisky," very human example, imagine seeing someone who piques sexual interest. Maybe it is love at first or second sight (or just lust). No matter, dopamine kicks in and makes the work of conquest feel effortless. Dopamine pushes one to do it. Over texting? Just do it. Dopamine would not steer one incorrectly – or would it? Constantly thinking about them? Spending too much money trying to impress? Impossible because it seems so right. Showing up at their workplace? Probably not a good idea, but dopamine does not allow one to thoroughly evaluate prior to one's acting out; rather, it is a chemical that pushes one to "do it" and "get it." Finally, the person of interest acquiesces. Upon successful conquest, dopamine release stops. After dopamine has done its job, now other neurotransmitters kick in.

Individuals who are addicted and/or dependent on a medication must have it and have it now, no matter the consequence. Dopamine is the neurotransmitter that makes the means to the end palpable, and the work of finding the medication (a "fix") easy. The addicted patient knows their actions are hardly rational. Still, withdrawal is so painful and the thrill of procuring the illicit medication is so exciting that the nucleus accumbens is in overdrive. The client is not really under the control of the medication; they are more likely under dopamine's spell. As a result, many drug abuse programs view drug addictions as a hijacking of the dopaminergic system.

Methamphetamine is an excellent example. Once upon a time it was a legal drug called Desoxyn®; however, its abuse potential continues to far outweigh its benefit. Not only is it full of dopamine, but by adding a methyl group (one carbon bonded to three hydrogen atoms, CH_3) to a common variety amphetamine the drug becomes much more lipophilic. It now crosses the Blood Brain Barrier (BBB) faster than a regular amphetamine. Its efficacy is magnified and causes the nucleus accumbens to respond with a rewarding high in a way that the brain is not evolutionarily wired to process. The high is beyond glorious, indescribable, never to be duplicated again. Because of its indescribable reward, the thrill of the dopaminergic chase is so powerful the addicted individual continues to chase the dream (incentive salience). As a result, substance abusing patients advise their therapists, "I must have it; and when I got it, it was not that good; but I had to have it." And that is the dopaminergic cycle of life.

Manipulating dopamine, either agonizing or antagonizing, is a staple of today's psychopharmacology. Doing so provides benefit to the patient but also a myriad of side effects. How do dopaminergic drugs do what they do? They manipulate the brain's four dopaminergic tracts or pathways. All four play a part in making dopamine the neurotransmitter that makes work seem effortless.

Never forget. Dopamine is the neurotransmitter that makes any work, activity, goal, product, need, want, desire, wish, even the pursuit of love seem effortless.

QUIZ TIME

(1) True or False? Dopamine is both a catecholamine and is a precursor of norepinephrine and epinephrine.

(2) A preponderance of dopamine is found in the _____.

      (a) substantia nigra

      (b) ventral tegmental area

      (c) basal forebrain

      (d) A & B

      (e) A, B, & C

(3) Dopamine makes work seem _____ and time _____.

      (a) effortless; speeds up

      (b) effortless; slows down

      (c) effortless; feels the same

      (d) effortful; speeds up

(4) Incentive salience means _____.

      (a) gratification can be delayed

      (b) desires and wants are motivational

      (c) behavior is amotivational

      (d) cognitive control of wants and desires

(5) Epinephrine is secreted from the adrenal _____.

      (a) gland

      (b) cortex

      (c) medulla

      (d) shaft

ANSWERS

(1) TRUE. Dopamine is a catecholamine, one of the two monoamines classes (the other is indolamine). Tyrosine (an amino acid created by the body) via enzymes turns into DOPA; DOPA via enzymatic action becomes dopamine. Then dopamine via enzymatic action is chemically altered into norepinephrine. Finally, norepinephrine via enzymatic action chemically transforms into epinephrine. Dopamine is a precursor for norepinephrine and epinephrine.

(2) The correct answer is "d," substantia nigra and ventral tegmental area. Not the basal forebrain. That is the home of acetylcholine.

(3) The correct answer is "a." Dopamine makes work seem effortless and creates the perception that time speeds up.

(4) The correct answer is "b," desires and wants are motivational. In the case of dopamine, it is a chemically motivated desire and want.

(5) The correct answer is "c." The adrenal gland secretes epinephrine; but more specifically, and a better answer, epinephrine is secreted by the adrenal medulla.

Module 25

One Neurotransmitter and Four "Tracts"

The Dopaminergic Pathways

☺

This module is all about one neurotransmitter — dopamine — and no train tracks, but "tracts." A tract is a group of neural fibers located inside the central nervous system; and for the sake of trivia, outside the central nervous system in the peripheral nervous system, while a tract is still a tract, instead of calling it a tract, it is called a ganglia. However, the early brain investigators were anything but consistent or clear in their initial naming. To explain, while the "basal ganglia" is part of central nervous system, it should, for consistency, be called the basal tract. It is, inconsistently, called the basal ganglia. To complicate matters, sometimes these dopamine tracts are referred to as "dopamine projections." For the sake of clarity, these tracts will be called dopamine pathways.

Dopamine pathways are neuronal connections that allow dopamine to travel to areas of the brain (and body) and facilitate (or impede) thoughts, reward, pleasure, voluntary movements (especially fine-motor movements), and executive functioning. Most psychopharmacology students study the four dopamine pathways. These four are integral to psychopharmacology; and medications agonize, partially agonize, antagonize, or inversely agonize these pathways. While there are eight (maybe more to be discovered, some sources say there are twelve) dopaminergic pathways, psychopharmacology concentrates on four.

The main four dopaminergic pathways are (1) mesolimbic, (2) mesocortical, (3) tuberoinfundibular (also called tuberohypophyseal), and (4) nigrostriatal. Two other pathways, not clearly understood, are the (5) medullary periventricular and (6) incertohypothalamic; however, it is thought these two may play a role in eating and circadian rhythms, respectively. The final two are (7) retinal and (8) glomerular pathways. As you would expect, dopamine in the retina visual functions. The dopamine glomerular pathway plays a role in regulating an interactive network of fluid homeostasis, blood pressure, and redox reaction balance (oxidation-reduction). Additionally, this dopamine pathway in the kidneys promotes sodium excretion and stimulates an anti-inflammatory pathway. These additional dopamine pathways may explain or predict some of the side effects of dopaminergic medications; however, it is the first four that are typically taught and considered essential knowledge for psychopharmacologists, prescribers, and all involved in integrated care responsible for protecting the patient (Choi, et al., 2015; Coffee & Brumback, 2006; Jackson, et al., 2012).

The Mesolimbic Pathway. Biological parlance is not as difficult as some people believe. The mesolimbic dopaminergic pathway is a neural pathway that carries dopamine, that much is clear from the last two words. Now dissect "mesolimbic." Chances are the word "limbic" is familiar. The limbic (from Latin, "limbus," meaning "border or edge") system is that part of your brain predominantly responsible for emotions. Put it together, and it translates to "emotional dopamine pathway."

"Meso" (from Greek, "mesos") means "middle or interior" brain and specifically refers to the mesencephalon part of the brain. The mesolimbic pathway is a neural dopaminergic tract that begins in the middle part of the brain (the mesencephalon) and runs to the brain's emotional middle part.

The mesolimbic dopaminergic pathway begins at the ventral tegmental (Latin "tegumentum" for "covering") area, part of the mesencephalon and continues to the nucleus accumbens (NAc) and amygdala (Latin "amygdalum," meaning "shaped like an almond"), both parts of the limbic system. The nucleus accumbens (NAc) is the reward center of your brain, and the amygdala is your danger detector. It warns by sending a signal to the hypothalamus which relays that signal to the sympathetic nervous system to prepare for fight or flight. Additionally, the amygdala attaches an emotional valence to memories and determines if a memory is a good, joyous, or a very bad, anger-provoking one.

The mesolimbic dopaminergic pathway is thought to be associated with positive symptoms of schizophrenia: Hallucinations and delusions. The theory developed by Arvid Carlsson in the 1960s purports medications that block this pathway decrease the intensity and/or make the hallucinations/delusions palatable. A substantial number of medications used today specifically block Carlsson's D_2 receptor (dopamine number two receptor).

The Mesocortical Pathway. This one is easier to verbally dissect. "Meso" means "middle," specifically the ventral tegmental area. "Cortical" is easy to figure out: Cortex. From the ventral tegmental area to the prefrontal cortex (PFC), this dopaminergic pathway is related to thinking and executive function. This area controls negative symptoms of schizophrenia. Often referred to as the "Six A's," they are (1) flat Affect, (2) Anhedonia, (3) inAttention, (4) Alogia, (5) Avolition, and (6) Associability. In other words, diminished emotional expression (flat affect), inability to feel pleasure (anhedonia), distractible (inattention), paucity of logic (alogia), lack of initiative (avolition), and unsociable (asocial). Regrettably, many current medications are not sufficiently effective in reducing schizophrenia's negative symptoms.

The Tuberoinfundibular Pathway. Concerning the tuberoinfundibular pathway, most do not have a pre-existing frame of reference to dissect this pathway. "Tubero" is from Latin "tubere," meaning "a swelling" (recall that a potato is a starchy "tuber"); "infundibular" is from Latin "infundibulum" meaning "cup" or "funnel." This pathway is also called the tuberohypophyseal pathway. The suffix is pronounced like "hypothesis," but with a "ph" instead of "th." Hypophysis cerebri is the old, original term for pituitary, and it comes from Greek meaning, "attachment underneath," and refers to the pituitary gland's position on the underside of the cerebrum. Putting all this together, the pathway starts at a swelling on the underside of the brain, the hypothalamus, and flows from the cup or funnel out of which the pituitary develops.

Because this pathway technically goes from the arcuate ("arch") nucleus of the hypothalamus to the median (middle) eminence (from Latin, "eminere," meaning "to project") of the pituitary, the median eminence connects or links the pituitary gland to the hypothalamus. Put all that together, this pathway is the link, arch or bridge from the hypothalamus to the cup or funnel which vascularizes the pituitary.

The tuberoinfundibular pathway is important because it inhibits the release of prolactin. This is a noteworthy side effect of many antipsychotic medications. Prolactin is a sex hormone released by the pituitary. It has a wide range of functions influencing behavior, regulating the immune system, and balancing sexual satisfaction.

Admittedly students grow very weary of their professor's continually saying, "Look at the word." Prolactin is a good example. "Pro" means increase; and the "lactin" is from Latin, "lacto," meaning milk. However, "increased milk" makes sense when examining dopaminergic medications that block the tuberoinfundibular pathway. A dopamine blocking drug stops this pathway's ability to inhibit prolactin's secretion thus indirectly increasing prolactin (hyperprolactinemia). As every late-night legal advertisement suggests, "If while taking this medication you developed gynecomastia or galactorrhea, call us." Translated: Numerous psychotropics indirectly increase prolactin levels. The results may be gynecomastia, enlargement or swelling of breast tissue in males; galactorrhea, milky nipple discharge; for females, breasts painful to the touch.

The Nigrostriatal Pathway. This pathway is critically important to remember; and to that point, an entire upcoming module is devoted to the side effects that occur when it is blocked. "Nigra" in this case refers to the substantia nigra, discussed in the previous dopamine module. The corpus striatum (Latin, for "striped body," named because the white matter of the internal capsule is overlaid and creates a striped appearance) is composed of the caudate and putamen. The caudate body or nuclei is located near the brain's center and rests on both sides of the thalamus; there is a caudate in both the left and right hemispheres. The caudate resembles a C-shaped structure with a wider head (Latin, "caput") in the front, tapering to a body (corpus) and a tail (Latin, "cauda") on both sides of the thalamus. The putamen, another part of the corpus striatum, derives its name from the Latin word for "shell" or "husk." The first investigators named so many parts of the brain for the things they resembled, the things that were part of their real world. Remember, too, that Latin was the language of science through the Middle Ages into the late 19th Century.

The caudate and putamen (both parts of the corpus striatum) are parts of the basal ganglia. When you think caudate and putamen, also think basal ganglia. Now, one additional word should be top of mind: Movement. The nigrostriatal pathway controls how well a person moves. Interestingly, this pathway contains around 80% of the dopamine in the brain (Coffee & Brumback, 2006).

Many psychotropics that either agonize or antagonize dopamine directly impact the nigrostriatal pathway. Too much stimulation (e.g., methamphetamine) or too much blocking (e.g., Thorazine®/chlorpromazine or Haldol®/haloperidol) can produce side effects which may include (but are certainly not limited to) tics, stereotypies, restless thoughts (thoughts are "movements"), spasms, tetany (involuntary contraction of muscles), tremors, motor restlessness, and extrapyramidal movements (e.g., Parkinsonian-like movements). A specific set of behaviors arise related with this pathway's being agonized or antagonized and are covered in a later module.

Since a substantial majority of psychotropics involve dopamine, these four pathways and their actions represent essential knowledge for any mental health professional. The mesolimbic pathway is linked with positive symptoms of schizophrenia. Mesocortical pathway and negative symptoms ("The Six A's") are connected. Hyperprolactinemia (increased amounts of prolactin) and concomitant side effects (i.e., gynecomastia and galactorrhea) are linked with blocking dopamine in the tuberoinfundibular pathway.

Last, but certainly not least is the relationship between movement disorders and the nigrostriatal pathway. When this pathway is either agonized or blocked, movement disorders can emerge. The four movement disorders are (1) dystonias, (2) akathisia, (3) neuroleptic-induced Parkinsonism, and (4) tardive dyskinesia. Knowing that these movement disorders result from these medications and monitoring for them can save many a patient from years of discomfort. More in Module 43.

QUIZ TIME

(1) Approximately _____% of dopamine in the entire brain is found in the nigrostriatal pathway.

    (a) 50%

    (b) 65%

    (c) 70%

    (d) 80%

(2) Which of the following is NOT one of "The Six A's"?

    (a) anhedonia

    (b) inattention

    (c) avolition

    (d) autism

(3) Which pathways are associated with positive and negative symptoms of schizophrenia, respectively?

    (a) mesocortical; mesolimbic

    (b) mesolimbic; mesocortical

    (c) tuberoinfundibular; mesolimbic

    (d) nigrostriatal; mesocortical

(4) Which side effect is associated with blocking the tuberoinfundibular pathway?

    (a) rhinorrhea

    (b) steatorrhea

    (c) diarrhea

    (d) galactorrhea

(5) Match the following:

_____Mesolimbic	A. Basal ganglia
_____Mesocortical	B. Visual functions
_____Tuberoinfundibular	C. Kidneys
_____Nigrostriatal	D. Negative symptoms
_____ Retinal	E. VTA to amygdala
_____Glomerular	F. Prolactin

ANSWERS

(1) The correct answer is "d," 80%.

(2) The correct answer is "d," autism. It is not one of "The Six A's" which include affect, anhedonia, inattention, alogia, avolition, and associability.

(3) The correct answer is "b," mesolimbic pathway is associated with positive symptoms (hallucinations and delusions), and mesocortical pathway is associated with negative symptoms of schizophrenia ("Six A's).

(4) The correct answer is "d," galactorrhea (milky nipple discharge). FYI: Look at the suffix, "rrhea." It is seen in numerous words including: rhinorrhea is a runny nose; steatorrhea is excreting large amounts of fecal fat; and diarrhea is diarrhea.

(5) Mesolimbic & "E," VTA/Amygdala; Mesocortical & "D," Negative symptoms; Tuberoinfundibular & "F," Prolactin; Nigrostriatal & "A," Basal ganglia; Retinal & "B," Visual functions; Glomerular & "C," Kidneys.

Module 26

In the Mood

Serotonin

☺ ☺ ☺ ☺ ☺

It is hard to believe but as late as 1937, many academics did not believe neurochemistry would become a viable discipline. It is even harder to believe that in the early 1950s, neurotransmitter theory was still controversial. At that time, only acetylcholine and norepinephrine had received any attention. It is even harder to believe that an unknown chemical which started out as an annoying contaminant, artifact, byproduct in the study of hypertension would change the world. It was just a thin, rhomboid, pale yellow crystal, that no one ever believed – except for one tenacious female researcher – would ever be found in the brain.

Who would have ever thought that a moody little inhibitor in the indolamine family would become the rock star of today? Admittedly it has taken serotonin over fifty years to become an "overnight success." Still, the discovery of serotonin was proof positive that the fields of neurochemistry and neurotransmitters were worthy fields of study. Then additional new discoveries of reuptake inhibitors and transporters, courtesy of work by Julius Axelrod and Georg Hertting, and a pharmaceutical marketing campaign turned serotonin into the ultimate neurotransmitter behind today's antidepressant medications (Axelrod, et al., 1961; Iversen, 2006, 2000). How all this happened is a wild ride of historical significance involving rabbits' gastric mucosa, octopus' saliva, and sixty 1.5-liter buckets of blood from the local Cleveland slaughterhouse gathered daily by one devoted researcher.

It all started in 1935 when Italian pharmacologist and chemist, Vittorio Erspamer discovered "something" in enterochromaffin cells (Renda, 2000). However, he did not realize the significance of his discovery and named it "enteramine," for two reasons. It was (1) an amine and (2) found in rabbits' enteric gastric mucosa. Enteramine was born. He also discovered it in octopus' saliva, and he named that derivative "octopamine" (Renda, 2000).

Erspamer was quite the savant. In the early 1930s, as a college student, he published his first work on biochemistry of the enterochromaffin cells. In 1935, he demonstrated that his "new discovery" made intestinal tissues contract. He believed he had discovered a new chemical; however, many of his peers questioned his discovery believing it nothing new, just a derivative of adrenaline. However, by its yellow color, he knew it was an indole. It took him two years, but his enteramine (1935) was indeed a new discovery followed years later by octopamine.

In 1948 he found octopamine in the salivary glands of the octopus, hence the name, octopamine. As a sympathomimetic, it mimicked norepinephrine and agonized the sympathetic nervous system. One thing that made Erspamer professionally unique, especially given today's interest in natural herbal remedies, was his unique devotion to developing pharmacologically rich substances from animals.

Octopuses aside, the key take away from Erspamer's work is that "enteramine" was found in the intestines and caused gut tissues to contract. It was a major source of study around 1952. Recall no neurotransmitter does only one thing. It was inevitable that this new enteramine would be found in other places in the body doing other things.

In 1948, researchers Maurice M. Rapport, Arda Green, and Irvine Page at Cleveland Clinic discovered a vasoconstrictor in blood serum. Diligently searching for the chemical source of hypertension, these researchers found that another substance was produced as soon as blood coagulated and had to be removed prior to synthesizing a hypertension stimulating factors (Whitaker-Azmitia, 1999). It was believed this intermediate substance had no research value, and no one at the time saw a relationship between this intermediate substance, serotonin, and mental health. It was just an intermediate substance, perhaps a contaminant, standing in the way of the chemical source of hypertension.

Isolating this substance required substantial quantities of blood. This vampire-like task of procuring blood for study fell to Rapport. For the next sixty mornings, on his way to the lab, he stopped by the local abattoir (slaughterhouse) and procured buckets of fresh blood. Finally, in 1951 this small intermediate substance was isolated (synthesized by a five-step chemical procedure) and produced thin, rhomboid, pale-yellow crystals. Noting it was found in blood serum, hence a "serum" agent, and that it affected vascular "tone," the name was obvious and inevitable (serum + tone): Serotonin.

For Rapport's hard and diligent work, the Research Committee of the Cleveland Clinic feted him with "a nice dinner party for his discovery" (Whitaker-Azmita, 1999).

That was the extent of the accolades; no one believed that "serotonin" had any value and was likely not in the brain. In 1952 these researchers definitively confirmed that enteramine and serotonin were identical substances. As a result, this was the death knell for Erspamer's enteramine.

Enter Betty Mack Twarog. Even though her mentor, Irvine Page, funded her search for brain serotonin, he advised that the search was likely futile because serotonin would <u>not</u> be found in the brain. In 1953, this talented researcher proved her supervisor wrong. She found serotonin in the brain which proved it was a neurotransmitter. Overnight, biochemistry and the science of neurotransmitters was born. This revelation would have been noted two years earlier in her previous research, but when Twarog submitted her results to the *Journal of Cellular and Comparative Physiology*, no one bothered to read it because it was written about an unknown neurotransmitter by an unknown female author (Page, 1968; Twarog, 1954).

While Twarog located serotonin in the brain's raphe nucleus, the role of serotonin in the brain was still unknown. The other researchers were myopically focused on the causes of hypertension. To their credit, they were instrumental in locating the peptide hormone, angiotensin (promotes aldosterone, a steroidal hormone which raises blood pressure), and its role in hypertension.

Serotonin's role in mental health would become the prescient achievement of a blind researcher named Dilworth Wayne Woolley. And of course, his opinions were initially rejected. Neurotransmitter theory was still just too new.

Woolley was a severe diabetic which eventually caused his blindness; however, he had the uncanny ability to "see" relationships where no relationship seemed obvious. Using a novel level of curiosity and unique creativity, Woolley's peers agreed he used all his remaining senses to find meaning in disparate things. In this case of relationships between disparate things, he wanted to know the connection between serotonin and lysergic acid diethylamide (LSD), first synthesized by Albert Hoffman in 1938.

Somehow "seeing" the chemical relationship between the structure of serotonin and LSD, he hypothesized that LSD was an "anti-metabolite" (he coined the word) of serotonin. An anti-metabolite is a substance that is structurally similar to another naturally occurring metabolite that could interfere, even block, the functioning of that natural substance or metabolite. He believed that LSD could interfere or block serotonin. In the 1950s, he submitted his hypothesis to *Lancet*. The editors rejected it and advised that when he proves his hypothesis, then they would reconsider publication.

In 1962, he published his hypothesis in *The Biological Basis of Psychoses or the Serotonin Hypothesis of Mental Health* (Wooley, 1962). Killed three years later while climbing in the Andes Mountains, it would take twenty-plus years before his work's value would be recognized, but he has never received the accolades he deserves.

Serotonin is a monoamine. Within that family, it is as Erspamer suspected by this substance's color, an indolamine. Approximately 90% of serotonin is found in the human gut; followed by 8% in blood serum; and only 2% is found in the brain (Yano, et al., 2015). While serotonin certainly has an important role in digestion and vasoconstriction, it is also known for other actions including its effects on:

- Mood (Jenkins, et al., 2016)

- Apathy (Barber, et al., 2018)

- Appetite (Anderberg, et al., 2017)

- Pain (Martin, et al., 2017)

- Panic (Maron & Shlik, 2006)

- Suicide (Underwood, M.D., 2018)

- Sleep (Oikonomou, et al., 2019)

- Intra-ocular pressure (Gunduz, et al., 2018).

- Osteoblasts in the Bone (Lavoie, et al., 2017)

- Areas not yet discovered

Who would believe a mild-mannered initially non-descript contaminating substance could turn a once upon a time, stepchild field of study, neuroscience, into a respected science? Who would have thought that a thin, rhomboid, pale, yellow crystal, turned neurotransmitter, would become the leading treatment for depression? Too bad the *Lancet* could not see what a blind man saw. Yet, that is precisely what happened.

QUIZ TIME

(1) Identify the genius.

    (a) Vittorio Erspamer

    (b) Maurice Rapport

    (c) Betty Twarog

    (d) Dilworth Wayne Wooley

    (e) All the above

(2) Serotonin is a(n) _____.

    (a) catecholamine

    (b) indolamine

    (c) peptide

    (d) neuraminidase

(3) Approximately what percentage of serotonin will be found in the gut?

    (a) 2%

    (b) 8%

    (c) 90%

    (d) 95%

(4) True or False? Neurotransmitters function very specifically and typically only do one thing.

(5) Grandma's medication is designed to increase acetylcholine. She needs it for her failing memory. Unfortunately, grandma is constantly complaining of _____.

    (a) salivation and lacrimation

    (b) urination and defecation

    (c) GI distress and emesis

    (d) slow heartbeat and bradycardia

    (e) all the above

BONUS QUESTION. Match the researcher to his or her finding, favorite research animal, fun fact, or anti-metabolite.

1. _____ H.I. Page   A. Discovered serotonin in brain

2. _____ D.W. Woolley   B. Rabbits' gastric mucosa & octopuses

3. _____ M. Rapport   C. Serotonin brain non-believer

4. _____ B.M. Twarog   D. Anti-metabolite

5. _____ V. Erspamer   E. Buckets of blood & free dinner

ANSWERS

(1) The correct answer is "e," all the above. No question. No doubt about it.

(2) The correct answer is "b," indolamine, just as Vittorio Erspamer suspected.

(3) The correct answer is "c," 90%. 2% of the body's serotonin will be found in the brain; and 8% will be found in blood serum.

(4) FALSE. All neurotransmitters, in conjunction with specific receptors, do more than one thing. While most psychopharmacology textbooks teach that serotonin is responsible for mood, they tend to omit that serotonin is implicated in mood, apathy, appetite, pain, panic, suicide, sleep, intraocular pressure, and osteoblasts in the bone.

(5) The correct answer is "e," all the above. Grandma is experiencing SLUDGEM: Salivation, lacrimation, urination, defecation, GI distress, emesis, and miosis. Since acetylcholine is the main neurotransmitter for the parasympathetic nervous system (PNS), and since this drug is overly agonizing it, it would not be unusual for the heart rate to slow (bradycardia).

BONUS QUESTION ANSWERS

1. Page & "C" Serotonin brain non-believer; 2. Woolley & "D" Anti-metabolite; 3. Rapport & "E" Buckets of blood & free dinner; 4. Twarog & "A" Discovered serotonin in brain; 5. Erspamer & "B" Rabbits' gastric mucosa & octopuses.

Module 27

Tug of War

Gamma Aminobutyric Acid (GABA) and Glutamate

☺ ☺ ☺ ☺

They are not monoamines. Certainly not catecholamines or indolamines. They are two of the most important amino acids in the field of psychopharmacology: Gamma Aminobutyric Acid (GABA) and glutamate (technically the anion of glutamic acid). Glutamate and GABA have a seesaw relationship, a tug-of-war relationship. Or to borrow an animal from Dr. Dolittle's menagerie (Lofting, 1920), these two neurotransmitters are the classic "Pushmi-Pullyu." Glutamate is the "Do It" to GABA's "Better not." Glutamate is a command, call to action; and GABA is the stand-down signal that restores calm. Every day, this balancing act plays out in the human brain.

Its presence known since 1883, GABA was originally thought to be a plant and microbe metabolite. That "nothing special" attitude appeared to follow the overall history of GABA's discovery. There are no rabbits, octopuses, or buckets of blood; and there is no major "Aha" behind the discovery of these neurotransmitters. Rather GABA's 1950 re-discovery by Eugene Roberts and Sam Frankel resulted from hard work and meticulously analyzing extract after extract. They identified it as a major amine in the brain and indicated that it preferentially accumulates there (Roberts & Frankel, 1950; Spiering, 2018). In other words, GABA calls the brain, "Home Sweet Home."

Roberts and Frankel's findings were greeted with a big "So what?" response from the scientific community. Remember, neurotransmitter theory was a debatable theory in the early 1950s. After GABA's discovery, no one had any idea of its function in the brain, and few peer-reviewed journals in the 1950s even talked about GABA. It took almost a decade after its discovery for researchers to realize that GABA inhibits neuronal action potential. Finally, in 1957 after an inhibiting amino acid was discovered in crayfish and identified as GABA, research on this inhibitory neurotransmitter started to take off. Roberts and Frankel's discovery was beginning to receive credit for a landmark breakthrough.

GABA (Gamma Aminobutyric Acid) is a common neurotransmitter that regulates neuroelectrical activity in the brain. The main GABA receptor is the GABA-A chloride ionophore – a fast receptor, preferential to the chloride ion. Via this receptor, many psychoactive drugs work to reduce anxiety, promote calm, and increase relaxation. Sleep medications are like a giant lullaby to the GABA-A ionophore. This ionophore also affects the body's daily circadian and infradian cycles (e.g., the body's menstrual cycles and seasonal rhythms).

Putting all this together, if a patient takes a drug like Valium®/diazepam (a benzodiazepine), an antianxiety or anxiolytic medication, it stimulates the GABA-A (notice that "A" is not written as a subscript) chloride ionophore. The result is reduced anxiety, enhanced relaxation, and increased sleep. Because numerous receptors in the brain's respiratory center are GABA-A chloride ionophores, this calm would be seen in more restful breathing. Additionally, there would be a reduction in the brain's neuronal action, so less possibility of seizures. By definition, medications that increase GABA are prophylactically anticonvulsant.

Because there are numerous GABA-A chloride ionophores in the hippocampus (part of the limbic system responsible for turning short-term memories into long-term memories), drugs that work here affect memory and make it more difficult to learn and to remember.

Finally, relaxation, calm, reduction in the brain's neuronal firing, and ataxia (a lack of voluntary movement control) is also possible. Any drug that affects the GABA-A chloride ionophore is a powerful medication with many positive responses (e.g., relaxation and calm) and numerous side effects (e.g., confusion, ataxia, and dependence).

On the other side of the inhibit-excite seesaw is glutamate. Not only is glutamate excitatory, but it is also the master switch in the brain. However, glutamate's initial history continues the story of skepticism. Decades after acetylcholine, norepinephrine, GABA, and even serotonin had been identified as neurotransmitters, glutamate, despite its history, was not generally accepted as a neurotransmitter. Scholars based their skepticism on specious reasons like (1) too much glutamate in the brain, (2) too little specificity, (3) lack of knowledge about its mechanism of action, and (4) no known antagonist. Finally, in the 1980s, after thirty years, researchers agreed that it was a neurotransmitter.

Glutamate started out in 1908 as a flavor enhancer. In fact, today more people are familiar with "monosodium glutamate (MSG)," a flavor enhancer in Chinese restaurants, than they are with "glutamate the neurotransmitter."

Biochemists looked for glutamate everywhere — in baby rats, cats, and toads; but Takashi Hayashi, at the Keio University School of Medicine in Tokyo, finally found its function in the cortex of dogs. The bulk of his glutamatergic research was performed during World War II; the English publication was delayed until 1954 (Hayashi, 1954). While high levels of glutamate were first found in the brain in the 1930s, speculation was rampant as to its role. Hayashi proposed that glutamate's function in the brain was excitation, causing neurons to be more active. He was right. However, interest in glutamate as a medication hardly flourished until the 1980s when glutamate was found to be a drug target offering treatment for epilepsy, spasticity, and neuroprotection (Watkins & Jane, 2006).

Glutamate can be your best friend or worst enemy. For glutamate to function as a neurotransmitter, it must be the right amount at the right place at the right time. Too much or too little can be problematic. It is the primary mediator of excitatory signals in the central nervous system and is integrally involved in most aspects of brain function. From cognition to memory to learning to drug cravings, it can be beneficial and turn quickly toxic.

Glutamate is a nonessential amino acid and can be easily synthesized in the body. Glutamate receptors consist of two families of eight receptors. The three ionotropic receptors (very fast) are NDMA, AMPA, and kainate; the remaining five receptors are metabotropic (fast, but slower than ionotropic). For psychopharmacologists, NDMA and AMPA are the most interesting. They represent about 70% of the glutaminergic receptors (Advokat, et al., 2019; Takagaki, 1996).

Because glutamate is toxic, it is only made when it is needed. Unused glutamate is broken down, removed quickly. Glutamate is removed via transporters into the nerve terminals and glial cells. Because the brain prefers inhibition (GABA-nergic action) to excitation (glutaminergic action), for every one molecule of glutamate that is broken down (by several mechanisms), it ultimately becomes two molecules of GABA. That means the ratio of inhibition to excitation is 2:1, GABA:Glutamate (Prauss, et al., 2014).

Patients who have exhibited traumatic head injuries are known for their impulsivity. That is likely because this 2:1 ratio, GABA:Glutamate, has been altered with an increase in glutamate signaling. To manage this problem, prescribers often try to restore this balance by prescribing GABA agonists or glutamate antagonists.

Push-pull, tug-of-war, seesaw, teeter-totter, yin-yang, id-ego, best friend-worst enemy, put on the brake-hit the gas, inhibit-excite — these opposites describe the inherent forces in the GABA and glutamate relationship. One thing is sure: Upset this delicate balance between these two amino acids, and problems will manifest quickly.

NEUROTRANSMITTER DISCOVERY TIMELINE

DATE	NEUROTRANSMITTER	DISCOVERER
1883	GABA	Unknown
1894	Epinephrine	Edward Albert Schafer & George Oliver
1901	Epinephrine	Purified by Takamine Jokichi
1907	Glutamate	Kikunae Ikeda
1910	Dopamine	Synthesized by George Barger & James Ewens
1910	Histamine	Henry Hallett Dale & Patrick Laidlaw
1913	Acetylcholine	Henry Hallett Dale
1921	Acetylcholine	Role identified by Otto Loewi
1935	Enteramine	Vittorio Erspamer
1946	Norepinephrine	Ulf von Euler
1948	Serotonin (5-HT)	Maurice M. Rapport, Arda Green, & Irvine Page
1950	GABA	Eugene Roberts & Sam Frankel
1952	Enteramine & Serotonin	Identified as identical neurotransmitters by Maurice M. Rapport, Arda Green, & Irvine Page
1953	Serotonin	Discovered in brain by Betty Twarog; finding published in 1954
1954	Norepinephrine	Identified in brain by Marthe Vogt

1954	Glutamate	Function identified by Takashi Hayashi
1958/ 1959	Dopamine	Function identified by Arvid Carlsson & Nils-Ake Hillard
1963	Biological Basis of Psychosis Hypothesis	Dillworth W. Woolley
1973- 1975	Endogenous Opioid Peptides/Ligands	John Hughes, Hans Kosterlitz, Eric J. Simon, Solomon Snyder, & Candace Pert
1992	Endogenous Cannabinoid Neurotransmitters	Lumir Hanus & William Devane

QUIZ TIME

(1) Which of the following is an inhibiting neuron.

(a) GABA

(b) Glutamate

(c) Epinephrine

(d) Norepinephrine

(2) The GABA-A chloride ionophore facilitates the passage of which ion (an atom or group of atoms that carries a positive or negative electric charge)?

(a) Potassium

(b) Sodium

(c) Acetylcholine

(d) Chloride

(3) GABA is _____; glutamate is _____.

(a) excitatory; inhibitory

(b) inhibitory; excitatory

(c) charged; protonated

(d) hydrophilic; water loving

(4) _____ is the "master switch" in the brain.

      (a) Acetylcholine

      (b) Norepinephrine

      (c) Glutamate

      (d) GABA

(5) Which of the following is NOT a glutamate ionotropic receptor?

      (a) Chloride ionophore

      (b) NMDA

      (c) AMPA

      (d) Kainate

ANSWERS

(1) The correct answer is "a," GABA. The other three are excitatory and/or activating.

(2) The correct answer is "d," chloride. Psychopharmacological and medical terms will likely advise what one wants to know in the "full name" of the receptor. In this case, the GABA-A chloride ionophore facilitates the passage of "chloride."

(3) The answer is "b." GABA is inhibitory; glutamate is excitatory – even toxic.

(4) The correct answer is "c." Glutamate is the "master switch" in the brain.

(5) The correct answer is "a." The chloride ionophore is a GABA receptor. The other three are ionotropic glutamate receptors.

Module 28

Greek to Me

Receptor Abbreviations, Symbols, and Letters

😊 😊

Many a psychopharmacology student has said to their professor (much the same way Casca said to Cassius in Shakespeare's *The Tragedy of Julius Caesar*, 1599), "It's Greek to me." Admittedly, receptor abbreviations are arbitrary, capricious, dated, and full of Greek and Latin. So here is a new analogy.

Music and medications have a lot in common. They both have specific identifiers that musicians and pharmacologists can "read" by looking at its notations. In music, there are twelve major and twelve minor keys. Psychopharmacology's receptors are like eight major keys — muscarinic, nicotinic, adrenergic, dopaminergic, serotoninergic, histaminergic, GABAnergic, and glutaminergic. Each of the eight receptor keys in pharmacology has its own letter, note, symbol, or abbreviation. Histamine is always abbreviated to "H." Dopamine is noted as "D." Serotonin is shorthanded to "5-HT." The parasympathetic nervous system designates its receptors as "M" and "N." Sympathetic nervous system (SNS) receptors use Greek letters like "α" (alpha) and "β" (beta).

In music, one key is selected for the musical rendition or passage. Two different musical keys cannot be played at the same time. Pharmacology is different. Any combination of all eight receptor keys can play in a drug at the same time.

Receptor abbreviations, symbols, and letters are not as complicated as musical notes, but they do require explanation. Like notes on a musical page, psychopharmacology's notes always refer to the same thing and create a picture of the drug, its action, and side effects. The question becomes, "If a drug's receptor notations are summarily listed, is it possible to 'sight read' and to know where a drug works and what the drug does?" Absolutely. Start by memorizing between ten to thirteen specific notations and both knowledge base and practice options increase dramatically. Here is a classic example.

Routinely used since 1961, Elavil®/amitriptyline was the second developed antidepressant (Tofranil®/imipramine was first). Knowing this drug's mechanism plus other impacted receptors is an excellent way to see a drug's multiple receptors, notable abbreviations, symbols, and letters.

Elavil®/amitriptyline is a drug in a class of medications called "Tri-Cyclic Antidepressants" or "TCAs." If this class of drug came to market today as a brand-new drug, it would be MOA classified as an "SNRI" – Serotonin Norepinephrine Reuptake Inhibitor. The receptors' notations are abbreviated using the letter "S" and "N." Prescribers use this drug to give the patient a mood and energy boost via serotonin and norepinephrine, but there are additional noteworthy receptors involved that are not delineated in its simple four-letter mechanism of action.

Elavil®/amitriptyline's recipe or full formula continues with the following receptor notations/ingredients. It is not solely a SRI and an NRI, it is also an α_1 (alpha one) antagonist, H_1 antagonist, M_1 antagonist, and $5\text{-}HT_{2A}$ antagonist and $5\text{-}HT_{2C}$

antagonist. Very few prescribers actively appreciate other specific receptor notations associated with this medication. Here is what you need to know to breakdown Elavil®/amitriptyline's notations. Here is the specific receptors' legend:

Elavil®/amitriptyline's Complete Drug Action[2]

(1) SRI

(2) NRI

(3) α_1 antagonist

(4) H_1 antagonist

(5) M_1 antagonist

(6) 5-HT_{2A} antagonist

(7) 5-HT_{2C} antagonist

The first thing to note is that most prescribers use this drug for its serotonin and norepinephrine reuptake inhibition, but Elavil®/amitriptyline is at least seven pills in one. The following is a breakdown of these receptors and their actions.

"RI" always means "reuptake inhibitor" — the drug blocks the reuptake transporter and does not allow neurotransmitters back into the terminal button. If it has an "S" in front of it that stands for serotonin. If it has an "N," in front of "RI," that stands for norepinephrine and will work in the sympathetic nervous system. If it has a "D" in front of the "RI," that stands for dopamine. If there is reuptake involved, then the capital letter of the neurotransmitter is used ("S" for serotonin; "N" for norepinephrine; "D" for dopamine). If the serotonin neurotransmitter is not involved

---

[2] More receptors are involved, but these are necessary to understand the abbreviations, symbols, and letters for most Tri-Cyclic Antidepressants.

in transport (reuptake) back into the terminal button, then 5-HT is used. A drug that is a "SNDRI" is a serotonin norepinephrine dopamine reuptake inhibitor. Like musical notes, that delineation never changes.

The lower-case Greek letter, α, (pronounced "alpha") advises that the sympathetic nervous system is involved. In Elavil®/amitriptyline's case, α_1 antagonist means that Elavil®/amitriptyline blocks the number one alpha receptor of the sympathetic nervous system. While it is not represented in the MOA of TCAs, a "ß" (beta) receptor is another sympathetic nervous system receptor.

If there is a "H" in the recipe, the drug either agonizes or antagonizes the histaminergic system. "H" relates to histamine. The number is the specific histamine receptor being stimulated or blocked. Elavil®/amitriptyline blocks the number one histamine receptor (H_1).

"M" denotes muscarinic receptors. This means that Elavil®/amitriptyline blocks the number one muscarinic receptor (M_1) of the parasympathetic nervous system. "M," muscarinic, refers to the parasympathetic nervous system.

5-HT (five-hydroxytryptamine) stands for serotonin. If reuptake transporters are involved, then "S" is used instead of 5-HT. If the notation refers to a specific serotonin receptor, then 5-HT is followed by that specific receptor letter/number (e.g., 5-HT_{2A}). In this case, Elavil®/amitriptyline is a SRI and also blocks 5-HT_{2A} and 5-HT_{2C} serotonin receptors (there are fourteen serotonin receptors).

"D" will always stand for dopamine; but there is no dopaminergic involvement with Elavil®/amitriptyline, so no "D" in the receptor recipe or formula. If dopamine was included, then the notation would note the specific receptor (e.g., D_2). There are five dopamine receptors.

Receptor notations advise Elavil®/amitriptyline is (1) a serotonin reuptake inhibitor and (2) a norepinephrine reuptake inhibitor. That is its mechanism of action, SNRI. Other receptors are involved without therapeutic benefit. These are (3) alpha one receptor blocker in the sympathetic nervous system, (4) histamine number one receptor blocker, (5) muscarinic number one receptor blocker (part of the parasympathetic nervous system), and then specifically blocks two of the fourteen serotonin receptors, (6) $5\text{-}HT_{2A}$ and (7) $5\text{-}HT_{2C}$. This is always written $5\text{-}HT_{2A}$ and $5\text{-}HT_{2C}$ – never "S_{2A}" or "S_{2C}." "S" is reserved, by convention, when serotonin is listed in the mechanism of action (e.g., SNRI) or to delineate a reuptake transporter (e.g., "SRI").

Elavil®/amitriptyline
MECHANISM OF ACTION:
SNRI "Seven Pills in One"

REUPTAKE/RECEPTOR NOTATION	EXPLANATION
SRI	Serotonin Reuptake Inhibitor
NRI	Norepinephrine Reuptake Inhibitor
α_1	Alpha 1, Sympathetic Nervous System
H_1	H, Histaminergic System
M_1	M, Muscarinic, Parasympathetic Nervous System
$5\text{-}HT_{2A}$	Serotonin Receptor 2A
$5\text{-}HT_{2C}$	Serotonin Receptor 2C

To reiterate:

- If a drug blocks the parasympathetic system, then it blocks one or more "M," muscarinic receptors.

- If a drug is anticholinergic, then it blocks the "M" muscarinic receptor, specifically "M_1."

- If a drug is an adrenergic blocker, it blocks "α" and/or "β" receptors.

- If a drug blocks reuptake or transporters, note the letter which precedes "RI." If it is "DRI," then it is a dopamine reuptake inhibitor. "S" for serotonin reuptake inhibitor; and "N" for norepinephrine reuptake inhibitor. You will never see "αRI" or "βRI."

- If a drug cites "5-HT," then a specific serotonin receptor is involved.

A drug's name, abbreviation, and/or symbol can tell one all about its side effects, too.

Imagine Ms. Jones – an imaginary patient – visits her prescriber complaining of sad mood, perhaps depressed, and experiencing little energy. The prescriber, in this case, prescribes "old yet modern" Elavil®/amitriptyline to improve her mood and energy level. That is the "SNRI" part. A drug was selected that has at least seven receptors involved, but only two are required for mood and energy. What about the other receptors that this drug stimulates or blocks? Here are a sampling of the notable side effects.

By blocking the alpha-one receptor ($α_1$), she might experience orthostatic hypotension; block the histamine number one receptor (H_1), sleepiness and weight gain will follow. The big one, muscarinic number one (M$_1$), is the receptor which when blocked provokes anticholinergic action (DUCTS — dry mouth, urinary retention, constipation, tachycardia, increased sedation, sweating, confusion, and visual problems).

RECEPTOR NAMES PER NEUROTRANSMITTER Common Receptors Integral to Psychopharmacology Legend	
NEUROTRANSMITTER	CORRESPONDING RECEPTOR NAME

Acetylcholine (ACh)	Muscarinic (M)
	
	 Nicotinic Receptors (Nicotinic-acetylcholine receptor, nAChR)
	   • N$_m$ also known as N$_1$, responsible for muscular contractions and found at the neuromuscular junction (NMJ)
	   • N$_n$, also known as N$_2$, found in the synapses between neurons.

Norepinephrine (NE)	Adrenergic Receptors
	   • Alpha (α)
	   • Beta (β)

Dopamine (DA)	Dopamine Receptors • D
Serotonin	Serotonin Receptors • 5-HT
Histamine	Histamine Receptors • H

Epinephrine	Epinephrine
	   • N$_m$ at NMJ
	   • Alpha (α)
	   • Beta (β)

GABA	GABA Receptors
	   • GABA-A
	   • GABA-B

Glutamate	Glutamate Receptors
	   • NMDA
	   • AMPA
	   • Kainate

Routinely prescribed Elavil®/amitriptyline is more complicated than it looks. It is not just two reuptake transporters (serotonin and norepinephrine), but also (3) alpha/SNS, (4) histamine, (5) muscarinic/PNS, and (6 & 7) two serotonin specific receptors are involved. Two receptors serve the therapeutic mechanism of action, and five cause unwanted side effects. It would have been wiser if Ms. Jones' prescriber had picked a cleaner, not so psychopharmacologically rich, more receptor-selective drug. Today's newer antidepressants are far more selective – but they still have side effects.

Einstein said, "Any fool can *know*. The point is to <u>understand</u>." To paraphrase, "Any practitioner can *know* a drug's mechanism of action. The point is to *understand* all their abbreviations, symbols, and letters." This knowledge base allows the clinician to match the correct diagnosis with the proper medication to the right receptor while minimizing side effects. Do it successfully, and it makes beautiful music.

QUIZ TIME

(1) True or False? Use 5-HT when referencing a specific serotonin receptor and the letter "S" when referencing serotonin as part of a drug's mechanism of action or reuptake inhibition.

(2) True or False? A drug's mechanism of action advises the prescriber of the drug's complete range of action.

(3) True or False? Blocking the M_1 (muscarinic) receptor yields some degree of dry mouth, urinary retention, constipation, tachycardia (rapid heartbeat), and increased sedation.

(4) True or False? Elavil®/amitriptyline is a relatively new drug, prescribed for a myriad of purposes (depression, headaches, pain, and others), inexpensive, and yields few side effects.

Match the neurotransmitter to its neurotransmitter name, symbol, or abbreviation – may have more than one.

(5) _____ Acetylcholine (ACh)   A. NMJ

(6) _____ Norepinephrine (NE)   B. Chloride ionophore

(7) _____ Dopamine (DA)   C. α, alpha receptors

(8) _____ Serotonin (5-HT)   D. NMDA

(9) _____ Epinephrine   E. M & N receptors

(10) _____ GABA-A   F. DA

(11) _____ Glutamate   G. 5-HT

ANSWERS

(1) TRUE. 5-HT is the specific receptor designate (e.g., $5\text{-}HT_{2A}$) and the capital letter "S" is used when serotonin is referenced in a drug's mechanism of action or reuptake/transport into the terminal button.

(2) FALSE. A drug's mechanism of action only advises how the drug is being used therapeutically. It does not advise all neurotransmitters involved. For instance, Elavil®/amitriptyline's mechanism of action is SNRI – serotonin norepinephrine reuptake inhibitor. It does not advise of the histaminergic, adrenergic, muscarinic, and additional 5-HT blocking action.

(3) TRUE. It is an anticholinergic drug that is blocking the M_1 (muscarinic) receptor. Anticholinergic side effects are dry mouth, urinary retention, constipation, tachycardia, and increased sedation (DUCTS).

(4) FALSE. Elavil®/amitriptyline is an old (1961) Tri-Cyclic Antidepressant (TCA). It is used for a myriad of purposes, many off-label. It is relatively inexpensive; but as it is prescribed more and more, supply and demand dictate the price will increase.

(5) Acetylcholine & "E" and "A," Muscarinic (M), Nicotinic (N), and neuromuscular junction (NMJ).

(6) Norepinephrine & "C," alpha (α) receptors.

(7) Dopamine & "F," both DA and the letter "D" are abbreviations for dopamine.

(8) Serotonin & "G," 5-HT (5-hydroxytriptamine, the formal name for serotonin).

(9) Epinephrine & "A," Neuromuscular Junction (NMJ); also "C," epinephrine works on alpha (α) and also on beta (ß) receptors.

(10) GABA & "B," GABA-A Chloride Ionophore (the complete name of the receptor).

(11) Glutamate & "D," NMDA – N-methyl-D-aspartate; also AMPA and kainate are the other two ionotropic glutamate receptors.

Permutations of Fun

Drug Ingredients

☺ ☺ ☺

With apologies to Mattel's Barbie™, "Mix and match it's fun to do, what drugs you make are up to you." To put a psychopharmacologic spin on this little ditty: Take the permutations of all the receptor letters, abbreviations, and symbols, throw them in a cocktail shaker, mix them all up, and creative mixtures emerge. For example, if a drug is a serotonin reuptake inhibitor, that translates to S + RI = SRI. A dopamine reuptake inhibitor follows the same notations that would be D + RI = DRI. If a drug is a known "SNRI" that breaks down to serotonin + norepinephrine + reuptake inhibitor.

The following is a list of standard medication notations. Write out what each one means, and then check the answers for complete explanation.

First, the easier ones guaranteed to know the last two letters.

(1) NRI: _____

(2) DRI: _____

(3) SNRI: _____

(4) NDRI: _____

(5) SNDRI: _____

Here are the answers:

(1) NRI is a norepinephrine reuptake inhibitor. Examples include Strattera®/atomoxetine, Edronax®/reboxetine, and Ludiomil®/maprotiline. However, here is the point. Strattera®/atomoxetine may not be familiar, but knowing that norepinephrine is one of the ingredients means that this is a very energizing medication. Remember, norepinephrine stimulates the sympathetic nervous system. As a result, this medication could be used to treat ADHD, narcolepsy, and weight loss. This, of course, is defined by FDA approval. So, if Edronax®/reboxetine (a drug sold in Canada and Europe) and Ludiomil®/maprotiline do not sound familiar, it is easy to ascertain from the drug MOA/legend what it could be used to treat.

(2) DRI is a dopamine reuptake inhibitor. Imagine a patient is given a drug like this one. What could it be used to treat? Think back to the module on dopamine. It is a rewarding neurotransmitter; it makes work seem effortless; it is a pleasure-awarding neurotransmitter; it is energizing, speeding up time. A good example of a hypothetical DRI is Provigil®/modafinil, and it is used to treat excessive sleepiness, fatigue, and depression.

(3) SNRI is a serotonin norepinephrine reuptake inhibitor. Among the many things that serotonin does, it is thought to elevate mood. Norepinephrine, which activates the sympathetic nervous system (SNS), is energizing. Elavil®/amitriptyline is a SNRI. It is an antidepressant based on the "S" and the "N." Examples here are Effexor®/venlafaxine, Pristiq®/desvenlafaxine, and Cymbalta®/duloxetine.

(4) NDRI is a norepinephrine dopamine reuptake inhibitor. An example is Wellbutrin®/bupropion. To determine what it is used to treat, look at the neurotransmitters. This drug blocks the return of norepinephrine and dopamine to the terminal button forcing it to stay in the cleft longer and do its job; therefore, it is energizing and pleasurable. Wellbutrin®/bupropion is approved to treat depression and seasonal affective disorder. Know the receptor shorthand and a logical idea of a diagnostic application follows.

(5) This abbreviation, SNDRI, puts it all together. It stands for serotonin norepinephrine dopamine reuptake inhibitor. Based on its pharmacology, it could elevate mood, provide energy, and make work seem effortless. A good example of a SNDRI is cocaine.

That is the idea. Look at the MOA nomenclature and figure out what the drug does. Try this next list. They are a little trickier, a little more thinking required, and here is a hint. The "A" will stand for either agonist or antagonist. Figuring out which one is the trick. Notice in the following questions, the "A" may or may not be capitalized. However, the "RI" never changes, so the last two letters will always stand for "reuptake inhibitor."

(6) SPARI: _____

(7) SaRI: _____

Here are the answers:

(6) The abbreviation, SPARI, stands for serotonin partial agonist reuptake inhibitor. It is serotonergic, so it can be used for mood. It is a partial agonist, meaning

that it binds with affinity and has an efficacy of about 20%. Notice that present convention dictates that a capital "A" implies agonist; and "RI" is, of course, reuptake inhibitor. What could this drug be used to treat? Depression? Anxiety? That is what the serotonin neurotransmitter is all about. A good example of a SPARI is Viibryd®/vilazodone. It is FDA approved for depression and is used off-label for anxiety. Look again at Viibryd®/vilazodone, a SPARI. This drug is a (1) reuptake inhibitor like Prozac®/fluoxetine and (2) a partial agonist of an important serotonin receptor, 5-HT$_{1A}$.

(7) The abbreviation, SaRI, stands for serotonin antagonist reuptake inhibitor. In this case, the capital "A" becomes a lower-case "a." Although there is no hard and fast rule, those in the know try to delineate agonist from antagonist by capital "A" and lower-case "a," respectively. A good example of a SaRI drug is Desyrel®/trazodone. This drug has two mechanisms of action. First, it is a reuptake inhibitor, meaning serotonin stays in the cleft longer. If serotonin is involved, then a good guess is that Desyrel®/trazodone is an antidepressant. Secondly, the "Sa" means that it blocks a specific serotonin receptor, in this case 5-HT$_{2A}$. More on this receptor later, blocking this specific one is a good thing since it is responsible for many possible side effects. Interestingly, Desyrel®/trazodone is rarely used for depression, it is more often used off-label for sleep.

Now, the next group. A little harder and requires out-of-the-box thinking and clearly demonstrates lack of clear logic in the MOA-naming nomenclature.

(8) NaSSa: _____

(9) DA$_2$ ant:_____

(10) SDA: _____

These three examples were chosen to illustrate that sometimes drug nomenclature does not follow any conventional rules, and that is a problem. There is an intervention afoot to remediate this problem; more on that at the end of this module.

(8) NaSSa, stands for noradrenergic antagonist serotonin specific antagonist or agent. This one is tricky. It blocks the autoreceptor of the sympathetic nervous system, and autoreceptors work backward. So, this drug blocks the receptor which inversely causes a substantial release of norepinephrine, providing energy. It is a serotonin-specific antagonist, meaning that it blocks a specific 5-HT receptor. In this case, 5-HT$_{2A}$, a serotonin receptor fraught with side effects. An excellent example of a NaSSa drug is Remeron®/mirtazapine. It is FDA approved to treat major depressive disorder.

(9) The abbreviation, D$_2$ ant, stands for dopamine number two receptor antagonist. Classic examples are Thorazine®/chlorpromazine and Haldol®/haloperidol. These are old, effective typical antipsychotics. They work by blocking the D$_2$ receptor.

(10) The abbreviation, SDA, stands for serotonin dopamine antagonist, even though the "A" is capitalized; too bad current drug-naming conventions are inconsistent. An excellent example of this class of medication is

Risperdal®/risperidone. It blocks both serotonin and dopamine. Recall that antipsychotic medications work by blocking dopamine. Risperdal®/risperidone is a textbook, atypical antipsychotic medication.

It is easy to reach one inescapable conclusion. This is confusing and difficult with minimal rational logic. To that point, there is a movement in play to standardize all drug receptor actions using "Neuroscience-based Nomenclature (NbN)." Starting in 2010, this initiative, promulgated by The European College of Neuropsychopharmacology (ECNP) and five other international organizations, today supports changes in the way professionals delineate psychotropic/psychiatric medications. This is essential in that the current discussion of receptor notations is based on a hodgepodge, alphabet soup of approved indications, chemical structures, mechanisms of action, and even creative marketing. Moreover, it is the bane of psychopharmacology students and professionals alike. To complicate matters, some international journals are currently insisting that medications use the new "Neuroscience-based Nomenclature (NbN)" as opposed to the standard that has been around for over fifty years.

The following table illustrates some of the previously cited medications using the old, but still used, nomenclature and its corresponding "Neuroscience-based Nomenclature (NbN)." The "Neuroscience-based Nomenclature (NbN)" app is available for free download in Android and iPhone App stores. Search "NbN2."

If this seems daunting, remember the following: "S" for serotonin, "N" for norepinephrine, "D" for dopamine, "H" for histamine, "ACh" for acetylcholine,

"AChEI" for acetylcholinesterase inhibitor, "RI" for reuptake inhibitor, "A" for agonist, and "a" for antagonist (admittedly not everyone follows this guideline). Attempting to mix and match them is a step toward mastery. Throughout the remainder of this programmed text, both standard Mechanism of Action (MOA) and "NbN" nomenclature will be used, if determined. Remember, "NbN" nomenclature is an evolving work in progress; not all medications have been newly "NbN" named.

DRUG RECEPTOR NOMENCLATURE		
Medication Name	Current Receptor Notations	Neuroscience-based Nomenclature (NbN)[3]
Strattera®/atomoxetine	NRI, norepinephrine reuptake inhibitor	N-RI, norepinephrine reuptake inhibitor
Provigil®/modafinil	DRI, dopamine reuptake inhibitor	D-RI, dopamine reuptake inhibitor
Elavil®/amitriptyline	TCA, Tri-Cyclic Antidepressant; SNRI, serotonin reuptake inhibitor	SN-MM, serotonin norepinephrine multimodal
Wellbutrin®/bupropion	NDRI, norepinephrine dopamine reuptake inhibitor	D-RIRe, dopamine reuptake inhibitor and releaser
Viibryd®/vilazodone	SPARI, serotonin partial agonist reuptake inhibitor	S-MM, serotonin multimodal

---

[3] Neuroscience-based Nomenclature (NbN) from *Stahl's Essential Psychopharmacology Prescriber's Guide* (6th Edition).

Desyrel®/trazodone	SaRI/SARI, serotonin 2 antagonist/reuptake inhibitor	S-MM, serotonin receptor antagonist, multimodal
Remeron®/mirtazapine	NaSSa/NaSSA, norepinephrine antagonist and serotonin specific antagonist	SN-Ran, serotonin norepinephrine receptor antagonist
Thorazine®/chlorpromazine	D2 ant, dopamine 2 antagonist	DS-Ran, dopamine and serotonin receptor antagonist
Risperdal®/risperidone	SDA, serotonin dopamine antagonist	DSN-Ran, dopamine, serotonin, norepinephrine receptor antagonist

QUIZ TIME

This module was a working quiz.

Fun with Drugs

Neurotransmitters

☺☺☺☺☺

Time to design a drug. After completing the previous modules, the knowledge foundation is in built. Time to put it together. Just answer a few questions. What will this drug do? How will it do it? What neurotransmitters are required to do the job? What problem are you trying to solve? And what problems are you creating? Working with "The Big Seven" neurotransmitters, here is how to put a drug together.

(1) Who would like a drug that makes you feel good?

(2) What about a drug that chills you out, takes away your anxiety, turns down the brightness of the day?

(3) What about a drug that allows you to forget your troubles, perhaps act as if they do not exist or just magically disappear? This drug would take away thoughts of financial worries, work problems, and/or family difficulties.

(4) Finally, what about a drug that would take away your pain?

Put all this together, and this medication would make you feel good, chill you out, make troubles vanish, and even take away your pain. Would it be therapeutic? Would it sell? Would this medication be popular? Would it be in demand? Here is how to make this miraculous medication. Ask and answer the following four questions.

Question 1: What neurotransmitter from "The Big Seven" is most likely to make you feel good? Which neurotransmitter makes work seem effortless? Which neurotransmitter is involved with one's reward system? Dopamine. So, to create this part of this special medication, there must be dopamine (D) in the mix.

Question 2: What neurotransmitter is clearly associated with reductions in anxiety? What receptor is an anxiolytic (antianxiety) ionophore? Which neurotransmitter reduces feeling like a rat on an electrified grid waiting to be shocked? The answer: GABA. To reduce anxiety, agonize the GABA-A chloride ionophore. Increasing GABA reduces anxiety.

Question 3: What neurotransmitter is associated with memory? Acetylcholine is certainly one possible answer. Agonizing acetylcholine improves memory and antagonizing it creates confusion; and there is no easy way to build acetylcholine in the system without SLUDGEM. So, is there another neurotransmitter choice which is associated with memory and forgetting? Is there one clearly associated with memory consolidation, long-term memory, and storage? Think glutamate. Blocking glutamate is associated with forgetting and memory loss.

Question 4: Which neurotransmitter selected will reduce pain? This neurotransmitter is foreshadowing a module to come. The best answer is an exogenous opiate/opioid to activate the pain-relieving mu (μ, little Greek letter) opiate receptor. This is one of four opioid receptors explained in a later module.

What about serotonin? Should it be included in this medication? It certainly has anxiety-reducing potential. Technically, serotonin "numbs," "dulls," "deadens," or "blunts." That is why serotonin is much more of a "serenic" than an antidepressant (some textbooks refer to "serenics" — from the word "serene" – and as anti-aggressives). In this case, leave it out, the other neurotransmitters are sufficient. In fact, in this case, given serotonin's side-effect profile, it might do more harm than good.

Mix and match these four neurotransmitters, and here is the product:

Stimulate Dopamine Receptor = Feel Good

Stimulate GABA Receptor = Chill Out

Block Glutamate Receptor = Forget Troubles

Stimulate Opioid Receptor = Pain Relief

Does this medication/drug have marketing potential? Absolutely.

Does a patient need a prescription for this medication/drug? No. Just over eighteen and in some states and twenty-one years of age in others.

Is there a co-pay with this medication? Not a co-pay but a tax.

Is this medication/drug self-reinforcing? If a person uses this medication, will they want to do it again? Yes. It has dopamine in it. It has opioids in it. It has GABA in it.

What is this medication? What is this drug? Alcohol (C_2H_5OH, ethanol, ETOH). Its mechanism of action (MOA) is (1) dopamine agonist, (2) GABA agonist, (3) glutamate antagonist, and (4) mu (μ) opiate agonist. Feels good, chills you out, forget troubles, reduces pain, no wonder it is called "Happy Hour."

Is it a perfect medication/drug? No. It takes more, and an explanation follows in the next module.

QUIZ TIME

(1) True or False? Histamine and serotonin are not part of alcohol's neurotransmitter recipe.

(2) True or False? Alcohol is self-reinforcing.

(3) True or False? Serotonin is an antidepressant that elevates mood.

(4) True or False? The mu (μ) opiate receptor is associated with pain relief.

(5) True or False? Thorazine®/chlorpromazine and Haldol®/haloperidol (typical antipsychotics) are D_2 antagonists.

ANSWERS

(1) FALSE. This is a trick question. Remember, the mechanism of action only advises why the drug is being used therapeutically. It does not inform of every neurotransmitter that may be involved in a drug's action. In this case, histamine and serotonin are both agonists in the overall, total notation of alcohol. Alcohol increases serotonin (Advokat, et al., 2019).

(2) TRUE. Alcohol is very self-reinforcing. Because alcohol stimulates the reward pathway, the probability that it will be used again increases. Because alcohol is an opioid agonist, withdrawal from the drug can be negatively reinforcing, increasing the probably that alcohol will be used again.

(3) FALSE. Serotonin is a "serenic." It numbs. By definition, an increase in serotonin decreases dopamine, it technically lacks in its overall ability to completely relieve depression. For marketing purposes, serotonin is the ultimate "antidepressant." In actual clinical practice, maybe not as much as the marketing suggests.

(4) TRUE. The mu (μ) receptor is currently believed to be the primary opiate receptor associated with pain relief. However, new discoveries (coming up later) may suggest the discovery of a new endogenous opiate receptor also equally effective for pain relief.

(5) TRUE. Thorazine® and Haldol® are textbook typical antipsychotics and D_2 antagonists. This is something all educated people know.

The Impossible Dream

The Perfect Pill

☺ ☺ ☺ ☺ ☺

Picking the right neurotransmitter is the first step in creating a medication, and it is likely the easiest step. However, it takes more than a neurotransmitter to make the perfect pill. What are the criteria for the perfect pill? It is a surprising, complicated, ever-changing, and perhaps impossible feat.

Spoiler Alert. There is no such thing as the perfect pill. Between pharmacokinetics (what the body does to the drug) and pharmacodynamics (what the drug does to the body), there is always room for improvement. Since every drug has side effects, every medication has imperfections. Nevertheless, all drug manufacturers strive to create the perfect pill. To design the perfect medication, the following qualities must be included. Think of them as the "Top 10," but there are likely hundreds of benchmarks necessary to make an impossibility even remotely possible.

(1) For the perfect pill to be perfect, it must have constant and complete efficacy – that is 100%. That means the designer wants the drug to work 100% of the time. It is a worthy goal but destined never to be achieved. For instance, antidepressant medications only work 60% of the time (not 100%) and are useful for approximately 60% (not 100%) of the population who takes the medication.

Regrettably, the remission rate for psychopharmacological drugs ranges from 20% to 40%. Current drugs offer neither constant nor complete efficacy (Meyer & Quenzer, 2019).

(2) The perfect medication would have a rapid onset of action. In other words, take the medication, and its healing qualities kick in fast. So many pharmaceutical commercials (more for over-the-counter medications) advertise that medications work quickly. However, in the real world, not really. Back to antidepressants, if they work, it can take up to four to six weeks for any "antidepressant" effect to kick in, and then only a 50% reduction in symptoms is considered a success. However, side effects kick in immediately.

(3) The half-life (t½) is essential in considering the perfect drug. Ideally, drug manufacturers prefer a twenty- to thirty-hour half-life. Since half-life determines dosing, a longer half-life medication increases the probability that the consumer will consistently take the medication. It is easier to remember to take one pill a day, and patients are more likely to take it. It requires work on the patient's part to remember to take a medication two, three, or four times a day. As a result, patient compliance decreases. Antidepressants are a good example here. Their half-lives run the gamut from nine hours to sixty-plus hours. Because the ideal drug has a long half-life, typically dosing once a day, most medications currently available hardly meet this standard of perfection (Fincham, 2005).

(4) The perfect medication has NO side effects. This is likely impossible to achieve. Admittedly some medications have fewer side effects; but *all* medications

have side effects. It is estimated that the antidepressant Prozac®/fluoxetine has hundreds of possible side effects. Some of them benign, mild to annoying; others harsh to deadly. The goal is perfection or at least minimal adverse drug reactions.

(5) Continuing with the goal of NO side effects, the perfect medication would be overdose proof with no toxicities. As Paracelsus pointed out in the 16th Century, it is the dose that alone determines toxicity. Perhaps an overdose free drug is a possibility in the future, but today the 16th Century admonition is still relevant.

(6) The perfect medication would have no active metabolites. An active metabolite emerges after the medication has been biotransformed in the body, usually by the liver. As a result, a healthy liver is a requirement. No healthy liver; no active metabolite; no drug effect. A drug that has an active metabolite will be less than effective in a patient with an impaired liver. Plus, the medication would tax the liver beyond its ability to perform. Ideally, take the medication; the liver does not need to metabolize it into anything and allows 100% (bioavailability) into the system, and the rest is healing. This rarely happens in the real world of today's pharmacology.

(7) On that same point, and referencing the liver once again, the ideal medication would have no drug-drug interactions. There would be no CYP450 inducers or inhibitors. Like children on a first-grade playground, all medications would "play well" together and not interfere with each other's actions. There would be no medication bullies. The perfect pill would be unique with no interplay with other drugs.

(8) If this were a perfect drug world, all medications would be "magic bullets" with multiple indications and applications. There is an interesting history behind the term "magic bullet" (also known as "silver bullet"). It was originally coined by German scientist, Nobel Prize Winner (1908) Paul Ehrlich, to describe antibodies and later the drug he developed, "salvarsan" (also known as "Arsphenamine" and "Compound 606"). Developed in 1910 as an effective treatment for syphilis, salvarsan was NOT a perfect drug, but, for the time, it served a purpose; moreover, this organo-arsenic compound (note the poison "arsenic" in this compound) was hailed as the first modern chemotherapeutic agent.

By definition, a "magic bullet" is a complete solution to any puzzling problem. It heals without any harm. A magic bullet is a drug that has multiple indications and applications and works equally well for each problem. It has no side effects. It has no toxicities. It eliminates the problem entirely with no harm to the host. It is perfect. It does not exist. The search for a magic bullet continues.

(9) Cost is another factor to consider in the search for the perfect pill. Prescribers rarely consider this, but the patient always evaluates it. The perfect pill would represent an affordable cost to all who need it. It would not be withheld for lack of payment but would be available to anyone regardless of ability to pay. Current costs range from free to exorbitant.

(10) The perfect pill would have the perfect name and a pleasing color; after all, no pill works if the patient does not take it. The more expensive the medication, the more attractive it must be. Usually, costly medications are varnished in ruby red,

emerald green, and sapphire blue tones. Medications that are no more than two syllables with "z" or "x" in the name (e.g., Prozac®/fluoxetine, Paxil®/paroxetine) are considered more "effective" (albeit "placebo effective") by the patient. Even a perfect pill will not work unless it is administered enterally (or parenterally). Taking the medication is a perfect pill's ultimate rate-limiting step.

Finally, and to paraphrase *Mary Poppins*' (1964) melody, a little bit of sugar does help the medicine go down, increase adherence, and improve sales. To that point, the perfect drug, pill, or syrup would have just a hint of sweetness. The sweeteners are typically sucrose, fructose, lactose, and synthetic aspartame and/or saccharin. Sweeteners can constitute half the drug weight in children's oral medications (Goldstein, 2015). Additionally, children's syrups can be infused with strawberry, banana, and raspberry flavorings for a nominal cost. The perfect medication would be pretty, sweet, and tasty.

To make the perfect pill, there are so many additional factors that must be considered including pharmacodynamics, affinity, intrinsic activity, tolerance, and many more. However, a goal for any medication must begin with a desire for perfection. Currently, psychopharmacological medications are far from perfect but represent a start.

QUIZ TIME

(1) True or False? It is easy to create the perfect pill.

(2) The scientist who coined the term, "magic bullet," was _____.

    (a) Otto Loewi

    (b) Sigmund Freud

    (c) Arvid Carlsson

    (d) Paul Ehrlich

(3) Of the following list, the most important factor in drug adherence is _____.

    (a) cost

    (b) color

    (c) half-life (t½)

    (d) potency

(4) True or False? All medications have side effects.

(5) Of the following list, which is the most important?

    (a) efficacy

    (b) potency

    (c) dissociation

    (d) first pass

ANSWERS

(1) FALSE. An easy question with a difficult outcome. Perhaps in the future the perfect pill will emerge. For today, there is no perfect medication.

(2) The correct answer is "d." Cigar smoking, hard drinking Paul Ehrlich created the term "magic bullet."

(3) The correct answer is "c," half-life (t ½). If the half-life is between 20-24 hours, the patient takes the drug once a day. Since that is easy to remember, it increases the patient's adherence to the medication regimen.

(4) TRUE. Side effects kick in with the first dose.

(5) The correct answer is "a," efficacy. If a drug does not work (efficacy), why take it?

Module 32

Pressing Buttons

The Terminal Button

☺

Where do drugs work? Sounds like a trick question and the answer is, of course, in the body. But more specifically, where do most of the drugs work their magic? The answer is at the site of the terminal button, presynaptic, and postsynaptic cleft.

The terminal buttons (aka "terminal bouton") are small bulblike structures located at the branching end of axons. Also called the "axon terminal," they represent the inflated portions (looks like little hands; each finger has a bulb-like protrusion or bulge) at the end of the axon. They contain vesicles (small sac-like structures) which hold and protect neurotransmitters. When action potential is achieved, along with an influx of calcium ions (exocytosis), vesicles release neurotransmitters which diffuse across the synaptic cleft. Presynaptic refers to a nerve cell's terminal button which releases neurotransmitters into a synapse following action potential. Postsynaptic occurs on the opposing side of the synapse. These extraordinary places yield many opportunities where drugs can exert their influence in at least nine ways.

DRUG ACTIONS AT TERMINAL BUTTON
1. Block Neurotransmitter Reuptake into Terminal Button
2. Agonize or Antagonize Receptors
3. Serve as Neurotransmitter Precursors
4. Inhibit Neurotransmitter Synthesis
5. Prevent Neurotransmitter Storage Vesicles
6. Stimulate Release of Neurotransmitters
7. Inhibit Release of Neurotransmitters
8. Stimulate Autoreceptors (Inhibits Neurotransmitter Release) or Block Autoreceptors (Increase Neurotransmitter Release)
9. Inhibit Neurotransmitter Degradation

(1) Block Neurotransmitter Reuptake into Terminal Button. Medications that are reuptake inhibitors (RI), drugs that block transport of neurotransmitters back into the terminal button, are common today. Most of the antidepressants used are serotonin reuptake inhibitors (e.g., Prozac®/fluoxetine), serotonin norepinephrine reuptake inhibitors (Effexor®/venlafaxine), and/or norepinephrine dopamine reuptake inhibitors (Wellbutrin®/bupropion). They block the designated neurotransmitter's path so this neurotransmitter must stay in the cleft longer increasing the probability of binding with its appropriate receptor. These medications, technically indirect agonists, represent approximately 33% of all psychotropic medications prescribed today. They block transporters causing neurotransmitters to remain in the cleft and ultimately agonize more receptors.

(2) Agonize or Antagonize Receptors. The most common drug mechanism of action is either to agonize or to antagonize postsynaptic receptors. Common examples are Thorazine®/chlorpromazine and Haldol®/haloperidol — old typical, antipsychotic

medications. Both medications block the D_2 (Dopamine number 2) receptor. The new atypical antipsychotics are also antagonists. They are Serotonin-Dopamine Antagonists (standard nomenclature refers to these as "SDA"). An example of an agonist would be Requip®/ropinirole, a dopamine agonist. Medications that are either agonists or antagonists represent approximately 50% of all psychotropics prescribed. If reuptake inhibitors represent 33% and agonists/antagonists represent 50%, the remaining 17% of medications work in novel ways in this terminal button-synaptic complex.

(3) <u>Neurotransmitter Precursors</u>. Drugs can serve as precursors which provide the ingredients to make neurotransmitters. Medications used to treat Parkinson's disease are an excellent example of precursor drugs. Since Parkinson's results from a depletion of dopamine, it makes sense that appropriate treatment would augment the patient's depleted levels by giving a patient a dopamine pill. However, dopamine is a charged molecule; it will never leave systemic circulation except to be cleared out of the system by the kidneys. To increase dopamine, prescribers use L-Dopa (a zwitterion, no net charge) that crosses the Blood Brain Barrier (BBB). In the brain, L-Dopa is the precursor (a substance that is converted into another one), and via enzymatic action (DOPA decarboxylase) is transformed into dopamine. The result, the precursor enters the system, provides the necessary ingredients, converts to dopamine, and serves as one treatment for Parkinson's by elevating the patient's dopamine levels.

(4) <u>Inhibit Neurotransmitter Synthesis</u>. Another way medications can work at the button-synaptic complex is by inhibiting or stopping a neurotransmitter's creation. Aspirin is a COX-2 inhibitor (cyclooxygenase inhibitor). This COX-2 enzyme facilitates

the creation of prostaglandins which have hormone-like effects and can function as a ligand. Prostaglandins cause pain, inflammation, and fever. Because it is inhibited, the COX-2 enzyme activates less prostaglandins, and pain perception and inflammation are reduced.

(5) <u>Prevent Neurotransmitter Storage in Vesicles</u>. Drugs can also prevent storage of neurotransmitters in vesicles. Ingrezza®/valbenazine is the perfect example. The FDA approved this drug in 2017 for the treatment of tardive dyskinesia (movement disorder explanations follow the antipsychotic modules). Blocking the Vesicular Monoamine Transporter Number 2 (VMAT2) reduces the vesicles' ability to load dopamine. Upon release of vesicles in the cleft, dopamine is reduced. It is theorized that reducing dopamine levels alleviates the symptoms (tardive dyskinesia) associated with dopamine hypersensitivity.

(6) <u>Stimulate Release of Neurotransmitters</u>. Amphetamines manipulate the terminal button complex by stimulating the release of neurotransmitters. Specifically, amphetamines have numerous mechanisms of action all designed to increase neurotransmitters in the cleft. This specific mechanism of action is explained in an upcoming stimulant module.

(7) <u>Inhibit Release of Neurotransmitters</u>. OnabotulinumtoxinA, better known as botulinum toxin (aka "BOTOX®"), one of the most poisonous substances known, exemplifies how a drug can inhibit a neurotransmitter release. BOTOX®/onabotulinumtoxinA blocks acetylcholine's release at the neuromuscular junction (NMJ). The result is flaccid paralysis (weakness, paralysis, and reduction in

muscle tone). Botulinum toxin is currently used to treat depression, heart issues, migraines, erectile dysfunction, back pain, sweaty palms, and 793 other problems (Sifferlin, 2017). Some of these 793 problems are approved by the FDA; most are not.

(8) <u>Stimulate/Block Autoreceptors</u>. Autoreceptors are tricky. They work opposite to an agonist or antagonist's expected action. When a medication agonizes an autoreceptor, it works like an antagonist and blocks it. When a medication antagonizes an autoreceptor, it works as an agonist and stimulates it.

Agonize Autoreceptor → Blocks Autoreceptor
Antagonize Autoreceptor → Stimulates Autoreceptor

Catapres®/clonidine and Tenex®/Intuniv®/guanfacine are two examples used to treat hypertension; Intuniv®/guanfacine is also approved to treat attention deficit hyperactivity disorder in children ages 6-17. The mechanism of action for these two drugs is an alpha-2 (α_2) adrenergic receptor agonist. The alpha-2 receptor is an autoreceptor that works like a thermostat (negative feedback loop). When it is stimulated, it blocks. When the heat goes up, the thermostat (autoreceptor) shuts things down. Putting this together, guanfacine stimulates the thermostat (autoreceptor) which turns down the heat (norepinephrine). Guanfacine agonizes the autoreceptor and blocks the sympathetic nervous system. As a hypertensive treatment, when the sympathetic nervous system is blocked, blood pressure lowers. When this drug is given to a hyperactive child, agonizing this autoreceptor, the child's energy level, via sympathetic nervous system blocking and decreasing norepinephrine, is reduced.

Remeron®/mirtazapine is a drug that works opposite Tenex®/Intuniv®/guanfacine. Mirtazapine blocks the alpha 2 (α_2) autoreceptor of the adrenergic, sympathetic nervous system. It blocks an autoreceptor, which stimulates that receptor. By blocking an autoreceptor of the sympathetic nervous system, it is stimulated. Now the sympathetic nervous system (The "5 F's") is turned on; norepinephrine is released; and the patient feels energized.

(9) Inhibit Neurotransmitter Degradation. Drugs can inhibit neurotransmitter degradation or breakdown. Classic, dangerous medications, the Monoamine Oxidase Inhibitors (MAOIs) are the best examples. The best known MAOI is Nardil®/phenelzine. These drugs block the enzyme that breaks down monoamines (remember "Cats live in the DEN"); as a result, dopamine, epinephrine, norepinephrine, along with serotonin (an indolamine) are not broken down and continue to build up in the cleft.

Another example involves Aricept®/donepezil, a drug used to treat memory loss. This medication is an acetylcholinesterase inhibitor (AChEI) and blocks acetylcholine's breakdown, increasing acetylcholine over the entire body, not just the brain where it is needed. Both MAOIs and AChEIs are indirect agonists.

Medications can agonize, antagonize, block reuptake/transport, inhibit neurotransmitter degradation, manipulate autoreceptors, block neurotransmitter synthesis, serve as precursors, and even enhance reuptake. All this happens in this tiny terminal button-synaptic complex. All action unknown to the patient – and numerous prescribers.

QUIZ TIME

Match the following:

(1) _____ Prozac®/fluoxetine   A. Negative feedback loop

(2) _____ Thorazine®/chlorpromazine   B. Classic and dangerous

(3) _____ RI   C. Acetylcholinesterase inhibitor

(4) _____ Agonists or antagonists   D. SSRI

(5) _____ L-Dopa   E. 33% MOA

(6) _____ Amphetamines   F. 50% MOA

(7) _____ BOTOX®/onabutulinumtoxinA   G. D_2 antagonist

(8) _____ Autoreceptor   H. Precursor

(9) _____ MAOIs   I. Releases neurotransmitters
                                               into cleft

(10) _____ Aricept®/donepezil   J. Blocks acetylcholine

ANSWERS

(1) Prozac, D – SSRI; (2) Thorazine, G — D_2 Antagonist; (3) RI, E – 33% of medications used are reuptake inhibitors; (4) Agonist, F – 50% of medications are agonists or antagonists; (5) L-Dopa, H – Precursor; (6) Amphetamines, I – Releases neurotransmitters into cleft; (7) BOTOX®/onabotulinumtoxinA, J – Blocks acetylcholine; (8) Autoreceptor, A – Negative feedback loop; (9) MAOIs, B – Classic and dangerous; (10) Aricept®/donepezil, C – Acetylcholinesterase inhibitor.

Relief Beliefs

Theories

☺ ☺ ☺

Somehow President George H.W. Bush's designating the 1990s as the "Decade of the Brain," morphed into "blame the brain," and finally this became the decade of "chemical imbalance." As a result, the common explanation said to many a patient ("You have a chemical imbalance in the brain, and that is why you are being prescribed this medication."), became ubiquitous and still in common use today. There is no evidence this explanation is true; to be brutally, ethically honest, there is no evidence that it is false. That said, presently, there is not one scintilla of evidence that a chemical imbalance in the brain is responsible for the patient's depression. It is all theory; and finally, much of the prescribing community has come to accept this chemical imbalance explanation as truth.

Theories have served as a "common explanation" in the mental health community for years. Unfortunately, this "chemical imbalance" explanation is theory only and any fact base is illusive. Nevertheless, although it has been hypothesis since the 1960s, this current viewpoint has been unwittingly promulgated as if it were true. Encouraged by pharmaceutical companies, too many prescribers jumped on this theoretical explanatory bandwagon as an easy, relief-belief way to justify the need for medication. At best, this explanation is theory; at worst, it is a violation of informed consent.

Thing chemical-imbalance hypothesis all started in 1958 when two groups of researchers presented independently of each other at an annual convention. The first group (Guy Everett and James Toman) was followed by the second group of presenters (John Saunders, Nathan Kline, and Maurice Vaisberg). Both groups offered papers, independently, during the 1958 meeting of the Society of Biological Psychiatry in San Francisco (published results came out in 1959). These presentations discussed "psychic energizers" – the previous name for antidepressants – and the newly emerging role of neurotransmitters. All researchers acknowledged during their presentation that this was an initiatory hypothesis (Everett & Toman, 1959; Saunders, et al., 1959).

Building on these earlier researchers' beliefs, Jacob Joseph Schildkraut (1965) introduced "The Catecholamine Hypothesis of Affective Disorders" or the "Biogenic Amine Hypothesis," which conjectured a biochemical cause and effect for depression. This hypothesis suggests that a scarcity of catecholamines (later expanded to include all monoamines — dopamine, norepinephrine, epinephrine, and serotonin) *may* be the basis of depressive disorders. Increase these neurotransmitters, as Schildkraut hypothesized, and depression will lift. It is important to note that Schildkraut also stated such a simplistic hypothesis could not possibly explain the source of depression, but he saw it as thought provoking, a place to start (Schildkraut, 1965).

For years, this hypothesis was nothing more than a one-paragraph mention or a footnote in a textbook. With the emergence of Prozac®/fluoxetine in the late 1980s, pharmaceutical companies dusted off Schildkraut's paper and added new wording from this old hypothesis to their current advertising campaigns. Amazingly, the original hypothesis made no mention of serotonin (because it is NOT a catecholamine), yet television and print commercials extolled "this old hypothesis" as the action by which these drugs "are thought to work" or "may" work. There was a legal and quasi-ethical brilliance in their judicious use of plausible disclaimers.

How "chemical imbalance" ultimately became viral, taught as fact, widely used, and overall accepted by all manner of clinicians is not known. Maybe it was a biological reason for what was thought to be a moral failing. In that regard, it served its purpose. However, what is known is that practitioners in all mental health specialties, especially psychiatry, vigorously embraced the "catecholamine chemical imbalance" for the last sixty-plus years. While its use has waned, even today it is not uncommon to hear, "You are depressed because you have a chemical imbalance in the brain." Precisely stated, this "chemical imbalance" concept is simplistic; it is inaccurate and cannot begin to explain the combination of biological and biochemical factors profoundly integrated in depression.

The catecholamine hypothesis of affective disorders was a hypothesis in 1965; it is still a theoretical hypothesis today. However, this old theory has morphed into the "Neurotrophic Hypothesis." Unable to die a merciful death, the old catecholamine hypothesis has just been updated. The newer theory makes no mention of "chemical imbalance," but focuses on a healthy neuronal growth protein that may serve as some sort of neuronal emollient.

The longer name for this new hypothesis is "Monoamine Hypothesis + Brain Derived Neurotrophic Factor (BDNF)." First, the word, "neurotrophic" comes from the Greek word, "trophikos," and relates to nutrition, specifically "to nourish." BDNF is a protein, encoded by the BDNF gene in humans. Foreshadowing, BDNF has many responsibilities including regulation of the immune system. But first, to the brain and depression. The "Monoamine Hypothesis + Brain Derived Neurotrophic Factor (BDNF)," also called the "Neurogenic Theory of Depression," is predicated on three recent discoveries: (1) Neuroplasticity — existing neurons can repair/re-model, (2) neurogenesis (roughly translated, nerve creation), and (3) discovery of BDNF (isolated in pig's brain and purified in 1982) by Yves-Alain Barde and Hans Thoenen, following the initial research of the prototypic "nerve growth factor" isolated in the 1950s by Rita Levi-Montalcini and Stanley Cohen.

BDNF is involved in the regulation, growth, maintenance, and proliferation of target neurons. In other words, it protects nerve cells and grows new ones. BDNF could easily stand for "Brain Derived Neurotrophic 'Fertilizer.'" It is "Miracle-Gro®" for the brain. It nourishes and promotes the health and stability of newly formed neurons. When new learning is involved (knowledge for which there is limited previous learning or no frame of reference), the brain grants a healthy dose of BDNF. Quality therapy causes the brain to produce BDNF. Problem solving? BDNF is your friend. Breathing exercises activate the Vagus Nerve (Cranial Nerve X, pronounced "Ten") which relaxes and increases BDNF (Holzel, et al., 2011).

Because psychotropic medications cause BDNF to be produced in the brain, it is hypothesized that BDNF is the mechanism, emollient, or treatment that causes depression to lift. Lower levels of BDNF have been observed in mental disorders including anorexia nervosa, depression, schizophrenia, and obsessive-compulsive disorders. The question is "Why do psychotropic medications increase BDNF in the brain?" Once again, theories and hypotheses provide possible explanations.

If there is no chemical imbalance in the brain, then psychotropics change brain homeostasis and unwittingly create a chemical change in the brain. The brain, recognizing this unexpected change in balance, seeks to protect itself from this insult. To achieve this protective end, BDNF is released. Concomitant depression lifts. So, it is not the modern medication that causes depression to lift. More accurately, it is the insult to the brain, and the brain's desire to protect itself that precipitates a release of BDNF which causes extra protection in the brain, additional neuronal growth, and a "soothing" effect. In other words, modern psychotropics inadvertently create iatrogenic harm. The brain attempts to resolve this chemical change, offers BDNF as an antidote. Depression lifts. Also, BDNF suppresses brain inflammation. As a result, a new theory has emerged. The keyword in the previous sentence is "theory."

This "insult to the brain" theory is so new, it does not have a name; but it is clearly gaining momentum. Perhaps stemming from an older theory which unites cortisol and the presence of depression, excess cortisol has also been shown to increase inflammation and decrease BDNF. Transitive property to the rescue, there must be a relationship between the three - depression, inflammation, and decreased BDNF. To explain, when inflammatory stressors are controlled, the negative feedback loop via the HPA Axis (hypothalamic-pituitary-adrenal axis) shuts down production of cortisol. Since there is an inverse relationship between cortisol and BDNF, BDNF will increase, and depressive symptoms will be reduced. Put another way: Since excess cortisol is related to decreased levels of BDNF, now, BDNF can return to its homeostatic balance. The new theory's central tenet is that inflammation and mental illness are somehow related. In fact, inflammation may be one of the precipitating precursors (Bullmore, 2018).

Why not take a pill for BDNF? It does not cross the Blood Brain Barrier (BBB). To increase BDNF, it must be done the old-fashioned way, through strenuous exercise, good nutrition, new knowledge (for which one has no frame of reference), good therapy, and mindfulness training/meditation. One important point: BDNF levels decrease with age. One must actively work to maintain healthy BDNF levels persistently throughout one's lifetime (Levitin, 2020).

The current modern and best psychological/psychiatric practice is based on old theories. Fifty years ago, researchers were asking, "How do antidepressants work?" Today, researchers, clinicians, and patients are still asking, "How do antidepressants work?" Today, there is a desire to move beyond theory to fact. Today, the desire to move beyond hypothesis and know the actual cause of mental illness continues to be researched – just as it has been studied for the last 75 years.

QUIZ TIME

(1) True or False? The original name for antidepressants was "psychic energizers."

(2) True or False? In 1965, the first hypothesis, Schildkraut's "Catecholamine Hypothesis of Affective Disorders" appeared. It is still taught today.

(3) True or False? Medications are necessary to relieve depression because the patient has a chemical imbalance in the brain.

(4) Today, the hypotheses that emerged in the early 1960s as a possible explanation for depression and mental illness in general _____.

    (a) have not been proven true

    (b) have not been proven false

    (c) are still theories today

    (d) all the above

(5) True or False? Serotonin is clearly an antidepressant neurotransmitter that elevates mood.

ANSWERS

(1) TRUE. Any time a medication has norepinephrine in its mechanism of action, it will psychically and physically energize.

(2) TRUE. Schildkraut would be proud. It is still taught today.

(3) FALSE. At this point in time, there is little to no evidence that depression is the result of a chemical imbalance in the brain. Only theories.

(4) The correct answer is "d," all the above. Theories then are still theories today. Although millions of dollars have been spent trying to verify these theories, to date they have neither been proven true nor false. Theories and hypotheses that were described in the 1960s are still precisely that: Theories and hypotheses.

(5) FALSE. It is at best a "serenic," (from the word "serene"). It numbs. It dulls. While it is called an antidepressant, in reality, technically, it is not.

Pious Frauds

Placebos

☺ ☺ ☺ ☺ ☺

DUMMIES! That was the original name for placebos. Neither nice nor politically correct, they referred to the drug's "counterfeit inaction" and not to the patient being "tricked" by them. Placebos work. Placebos work well. Placebos may account for 75% of the antidepressant response (Kirsch, 2010). Ethical or unethical? That is a question for another day. How common is their use? As many as 45% of prescribers admit to using them (Reeves, 2007). In 1807, Thomas Jefferson, referring to placebos as "pious frauds," said, "One of the most successful physicians I have ever known has assured me that he used more bread pills, drops of coloured [sic] water, and powders of hickory ashes, than of all other medicines put together" (de Craen, et al., 1999). However, once one understands the role of placebos, it is evident that 100% of all prescribers indirectly and/or directly use "dummies."

Placebo is Latin for "I will please" and refers to pleasant effects of treatment. If a placebo goes awry, it is referred to as a "nocebo," Latin for "harm."

The official creation of placebos and the "art of deception" as a medicinal aid started in World War II's Italian Campaign (Allied operations in and around Italy between 1943 and 1945). Approximately 60,000 Allied soldiers died and about 320,000 were wounded. As a result, there was a substantial shortage of supplies, especially medications for pain. The hero of the story is not the physician, Henry Beecher (1955) who took credit for this "deception," but his nursing assistant. She is unknown to history, not even a footnote in Beecher's printed works.

It appears that this nurse, no doubt a result of her desire to mitigate her patients' overwhelming discomfort, injected soldiers with saline and assured the patient that the infusion was a potent pain reliever. The result of this inert dose was (1) a decrease in agony and (2) a decrease in blood pressure which prevented life-threatening shock (Beecher, 1955).

This type of placebo, injection of normal saline and "dummy" sugar pills are examples of "pure placebos." Examples of impure placebos are off-label use of medications (e.g., antibiotics to treat a virus), lab tests and physical exams for assurances, inactive medical devices, and surgeries for appearance (e.g., a small incision only with no further procedure). Pill color, size, and price are also placebos. Red is more stimulating; blue is more calming. The bigger the pill, the bigger the placebo effect. The greater the price, the longer lasting the effect.

There is also a newer use for placebos called "active placebos." They are placebos that are not "total dummies" and cause a mild side effect or two. Active placebos are employed in some double-blind drug studies. Imagine a drug researcher investigating a potential medication. In phase studies, the new investigational drug is compared to placebo with participants randomly divided into two groups. One group receives the new investigational medication, and the other receives sugar-pill placebos. When one subject group receives the study drug and side effects appear, and no side effects appear in the second group, the researcher has unintentionally unmasked which participants are receiving the actual medication and which are receiving placebo. If, instead of a pure placebo, one group is given an "active placebo" and the other receives the study medication, both groups will present with side effects making it more difficult to "accidentally" unmask or unblind the study, thus making the study more "blind." The ethics are questionable, debatable, and often overlooked.

To that point, if 75% of antidepressant effectiveness is the result of placebo effects, some hypothesize that the remaining 25% of antidepressant effects are due to active placebo (Kirsch, 2010; Kirsch & Sapirstein, 1998).

Placebos are not effective with every patient; however, if the patient exhibits a desire to please, a simple placebo may heal. Not all placebos look like medications. A simple smile, encouragement, patient communication, clear instructions, an expression of patient interest, graciousness, and a simple "thank you" are all placebos and have an effect in every practice facility.

QUIZ TIME

(1) Which of the following is/are types of placebos?

    (a) active

    (b) pure and impure

    (c) pill names

    (d) pill color and size

    (e) all the above

(2) Placebo is Latin for "I will _____."

    (a) heal

    (b) please

    (c) fix

    (d) harm

(3) What percentage of antidepressant effect is theorized to be the result of placebo?

    (a) 25%

    (b) 50%

    (c) 75%

    (d) 99.9%

(4) True or False? Placebos work on everyone for every condition.

(5) Of the following list, the most important factor in drug adherence is _____.

    (a) cost

    (b) color

    (c) half-life (t½)

    (d) potency

ANSWERS

(1) The correct answer is "e," all the above, along with positive reinforcers like smiles, encouragement, listening, clear instructions, graciousness, and a simple "thank you."

(2) The answer is "b." A placebo is Latin for "I will please." A nocebo translates to "I will harm."

(3) According to Kirsch (2010), the answer is "c," 75% of an antidepressant's clinical effect is thought to be the direct result of placebo.

(4) FALSE. This is a "too good to be true" question, and when they look "took good to be true," they are always false. Placebos work best with patients who have a desire to please the practitioner. They will not work in every condition.

(5) The correct answer is "c," half-life (t½). A half-life around 24 hours is ideal because the medication's dosage is once a day. Patients are likely to adhere to a once-a-day schedule. However, it would be easy to make an acceptable argument for cost. Give full credit for "a" or "c."

To Prescribe or Not to Prescribe

Therapy vs Antidepressants

😊 😊 😊 😊 😊

When is it appropriate to prescribe an antidepressant? When is it appropriate to provide therapy? When should one advocate for both? There actually is an acceptable answer, even a rule, that answers this question.

If one has a hammer, the world is a nail. If, however, one's specialty tool is a level, then the world is out of balance. The same is true with medications. Prescribers who are trained to prescribe, do just that, prescribe medications with an occasional nod toward talk therapy interventions. On the other hand, clinicians trained in therapy, view talk therapy as the panacea over chemical interventions.

Here are the age-old questions: (1) When is "talk" therapy recommended? (2) When are medications recommended? (3) When does one recommend both, therapy and drug? (4) Is there a recommendation for drug/medication only for the treatment of depression?

The answer: If the patient is experiencing mild to moderate depression, therapy is the recommendation. If the patient is experiencing moderate to severe depression, medication plus talk therapy, usually Cognitive Behavior Therapy (CBT) wins out. One would be hard-pressed to find a situation in the research literature where drug therapy alone is the treatment of choice for depression (Fournier, et al., 2010).

Assume a patient is requesting services for depression. During the clinical interview, they advise they are suffering from insomnia. Not sleeping and depressed are the two main clinical features associated with this patient. In this situation, it would be difficult to assess the severity of this patient's depression accurately. Any attempt at therapy would be met with a big yawn from the patient along with an inability to concentrate. What to do? Therapy or drug? Or both?

In this case, getting the patient restful sleep is the first helping task. Therapy could focus on sleep hygiene, but they need to rest immediately. Is there a medication, which at a low dose will help them sleep? The antidepressant, Remeron®/mirtazapine (MOA: NaSSa; NbN: SN-RAn) at a low dose (15 mg) fits the bill. At this low dose, it is more of a treatment for insomnia than for depression. Treatment for depression takes a back seat until its severity can be accurately assessed.

Back for a medication review, and the patient is sleeping well now. The medication worked. Now, use an assessment tool (e.g., Hamilton Depression Rating Scale) and evaluate their level of depression. If a pre-test assessment had been administered, successfully managing their sleep would see a posttest score change of several points with a reduction in depression severity. Now the question becomes: Is it mild to moderate or moderate to severe depression?

If the assessed depression is mild to moderate, therapy only is the recommendation. Sleep hygiene techniques to improve his sleep architecture is a worthy treatment goal. Plus, cognitive therapy of some type helps identify the depression's source, manage the difficulties, and ultimately resolve the problem. Additionally, it is also appropriate to taper the medication's use. While this drug, Remeron®/mirtazapine, is being used to treat insomnia, this is an off-label use of this medication. At lower doses, it has a powerful antihistaminergic quality which yields a treatment for sleep but an unwanted increase in appetite.

If, upon drug review and another assessment of depression, it is adjudged that the depression is moderate to severe, then drug plus therapy is in order. The dose of Remeron®/mirtazapine can be increased, and at higher doses (greater than 15 mg), it becomes more of a true antidepressant. Remeron®/mirtazapine is approved by the FDA for the treatment of major depressive disorder.

Is there ever a time when drug only for depression is the treatment of choice. The literature does not support a "drug-only, with no therapy approach." Perhaps there is an exception somewhere or a stipulation/demand from an insurance carrier, but drug only is not a recommended best practice treatment for depression. Therapy plus medication is always the preferred choice. Therapy teaches the patient how to resolve the difficulties intrinsically; medication is an extrinsic treatment that requires little intrinsic participation from the patient.

When it comes to the number of medications prescribed for a patient, there is another goal in mind. If the patient is depressed, the goal is one treatment antidepressant. If the patient is psychotic, the goal is one antipsychotic. If the patient is bipolar, typically more than one, perhaps two medications are recommended. In all situations, some form of therapy is a correct recommendation (Heldt, 2017).

Three Greek aphorisms, inscribed on the pronaos (portico) of the Temple of Apollo at Delphi, summarize both the tenets and ethics of this module. First, "Surety brings ruin." If either a prescriber or therapist believes their way is the best, the only way, it is problematic for the patient. Second, "Nothing to Excess," both the need for therapy and medication can be used to excess and must continuously be evaluated – at every session. Mild to moderate depression, therapy only. Moderate to severe depression, therapy plus medication. Finally, ethics dictate that the healer must "Know Thyself" and thy opinion on the value of medications versus therapy, or treatment is contaminated (American Psychological Association, 2011, 2002).

QUIZ TIME

(1) True or False? The Hamilton Depression Rating Sale is one of many available scales that can be used to assess a patient's level of depression.

(2) True or False? It is more challenging to assess a patient's level of depression when chronic insomnia is one of the main complaints.

(3) True or False? If a patient is experiencing mild to moderate depression, therapy only is likely the treatment of choice.

(4) True or False? If a patient is experiencing moderate to severe depression, therapy plus antidepressant medication is likely the treatment of choice.

(5) True or False? No drug or therapy works without the correct diagnosis.

ANSWERS

All quiz questions are TRUE.

Module 36

Mood Boosters

The Monoamine Oxidase Inhibitors (MAOIs)

☺ ☺ ☺

The first, true antidepressants emerged in the early 1950s and were discovered by accident. Monoamine Oxidase Inhibitors (MAOI, and pronounced sounding out each letter, M-A-O-I) were discovered by scientists working with tuberculosis patients in local sanitariums which were still around in the 1950s (even though the development of streptomycin during World War I had brought an end to many convalescent centers in the United States). These practitioners realized that tuberculosis patients taking the medication, iproniazid, a hydrazine derivative of isoniazid, noted visible positive changes in mood and alertness. Even though many of the patients had damaged lungs, there was still laughter and visible happiness. As a result, in the mid-1950s, iproniazid (a non-selective, irreversible MAOI) was approved as an antidepressant medication. While this drug has been discontinued (due to possible liver damage), other MAOIs have also found disfavor due to dangerous side effects. Today, these medications are used sparingly. However, these early MAOI antidepressants spurred research to create new formulations and antidepressant medications that are on the market today.

MONOAMINE OXIDASE INHIBITORS (MAOI)

BRAND (Year Entered US Market)	GENERIC	MOA[4]/NbN[5]
Nardil® (1961)	phenelzine	MAOI/ SN-EI, serotonin norepinephrine enzyme inhibitor; the NbN-2 app includes dopamine
Parnate® (1961)	tranylcypromine	MAOI/ SN-MM, serotonin norepinephrine dopamine multimodal enzyme inhibitor
Marplan® (1959)	isocarboxazid	MAOI/ SN-EI, serotonin norepinephrine enzyme inhibitor
Emsam® (1989)	selegiline	MAOI (Selective MAO-B inhibitor)/ DSM-EI, dopamine serotonin norepinephrine enzyme inhibitor

Of the main three MAOIs, Nardil®/phenelzine is considered the "gold standard," but all MAOIs work by blocking monoamine oxidase. As a result, monoamines are "boosted," and all monoamines (including catecholamines and indolamines) build up in the synaptic cleft. MAOIs are indirect agonists. It was initially thought (courtesy of the "Biogenic Amine Theory") that this "boost" would translate into a "mood boost." MAOIs do cause an almost immediate increase in neurotransmitters; but any antidepressant effect, even though there is a quick boost in serotonin, norepinephrine, and dopamine can take weeks — if at all.

---

[4] Mechanism of Action (MOA)/not to be confused with Monoamine Oxidase (MAO), the enzyme
[5] Neuroscience-based Nomenclature from *Stahl's Essential Psychopharmacology Prescriber's Guide*, 6th Edition.

The more common side effects from MAOIs include (note the mnemonic):

- **M** – Myoclonus → Quick, involuntary muscle jerks
- **A** – Anticholinergic → DUCTS → Dry mouth, urinary retention, constipation, tachycardia, and sedation
- **O**rthostatic Hypotension → (aka "postural hypotension") Dizziness upon standing
- **I** – Intracranial Hemorrhage → Intracranial brain bleed from ruptured blood vessels
- **S** – Sexual → Loss of libido

Because MAOIs cause norepinephrine (NE) levels to increase in the peripheral nerves of the sympathetic branch of the autonomic nervous system, even over-the-counter drugs (e.g., some nose sprays, common cold treatments, asthma medications) can have a potentiated effect (to make one of the medications more potent). There is one other monoamine (in addition to dopamine, epinephrine, norepinephrine, serotonin, and melatonin), tyramine, which builds up when monoamine oxidase enzymes are blocked. Tyramine is a big problem.

Tyramine increases when monoamine oxidase is blocked, and the elevation can be life threatening. Tyramine, a naturally occurring amine, is formed as a fermentation byproduct of digesting foods. Tyramine's harmful capability is easy to remember because it sounds like "mean." MAO-A (Monoamine Oxidase A) is responsible for breaking down tyramine, and MAO-B serves another purpose and is discussed later. With the proper amount of enzyme, monoamine oxidase breaks down tyramine; however, if the enzyme designed to break down this "mean amine," tyramine, is reduced, then tyramine becomes a rogue sympathomimetic, amplifying the powers of the sympathetic nervous system.

This augmentation happens when tyramine is absorbed into the bloodstream. Here it mimics the action of a sympathetic nervous system agonist's action causing blood pressure to ratchet up unabated, heart rate to increase dramatically, rapid breathing, and the possibility of a thunderclap headache, which foreshadows a deadly intracranial hemorrhage. Naturally occurring tyramine is acceptable given the proper enzyme levels to break it down. However, certain foods are known to increase tyramine. As a result, patients taking MAOIs must be careful which foods they eat. If they do not, then tyramine becomes potentially deadly.

This is called the "wine and cheese effect." Mix certain types of wines and cheeses with a monoamine oxidase inhibitor, and tyramine (a sympathomimetic) dramatically increases unabated in the system leading to deadly side effects. The following is a partial list of foods that may **NOT** be eaten while taking a monoamine oxidase inhibitor (MAOI):

- Aged cheeses
- Tap (draft beer), unpasteurized beer, and certain wines
- Cured meats (e.g., sausage, pepperoni, and salami)
- SPAM®
- Soy Sauce
- Yeast-extract spreads (e.g., Marmite®)
- Sauerkraut
- Bananas
- Red plums
- Avocados
- Raspberries
- Champagne
- Caviar

Emsam®/selegiline is a monoamine oxidase inhibitor (NbN: DSM-EI, Dopamine, Serotonin, Norepinephrine Enzyme Inhibitor) with a dietary twist. It is MAO-B preferential, and this allows the drug to be used with minimal dietary restrictions.

To explain, there are two types of monoamine oxidase, Type A and Type B. Both types are found in neurons and astroglia (shaped liked "stars," non-neuronal cells in the brain and spinal cord, responsible for structure, support, waste removal, neuronal protection). MAO-A metabolizes serotonin, norepinephrine, and dopamine; additionally, MAO-A is abundant in the gastrointestinal tract. MAO-B metabolizes dopamine. Because selegiline is preferential to MAO-B, this drug is virtually dopaminergic, albeit mildly. As opposed to a noradrenergic (norepinephrine) medication, and because of minimal impact on norepinephrine, the "wine and cheese effect" is not as pronounced with Emsam®/selegiline. Nevertheless, dietary intake and possible side effects should be monitored. Unfortunately, making this medication more "food friendly" reduces clinical efficacy.

One final point about the MAOIs: They do not mix well with other drugs. As a result, at least two (or more) weeks must be allowed for MAOIs to washout of the system prior to starting other antidepressants. Starting another antidepressant too soon after finishing a regimen of MAOIs may lead to "5-HT Syndrome or Serotonin Storm" (covered in an upcoming module). During this washout period (minimum two weeks), a patient should continue food and beverage restrictions. MAOIs have specific drug interactions with cough syrups with dextromethorphan (DM), headache remedies, painkillers, SRIs, and decongestants like pseudoephedrine.

Given the peculiar idiosyncrasies of MAOIs, problematic side effects, and food-unfriendly nature, these medications tend to be the antidepressant of last resort. Ironically, the first antidepressant created is now the last drug of choice.

QUIZ TIME

(1) Which of the following neurotransmitters is NOT a monoamine?

    (a) Glutamate

    (b) Serotonin

    (c) Norepinephrine

    (d) Dopamine

(2) The purpose of the new and improved "Neuroscience-based Nomenclature (NbN)" is to

    (a) make it more indication, rather than disease driven

    (b) reduce polypharmacy, using more than one medication

    (c) create clearer terminology

    (d) all the above

(3) The creation of MAOIs was _____ and discovered in the 1950's tuberculosis sanitariums.

    (a) accidental

    (b) empirically based

    (c) FDA approved

    (d) derived from the Tri-Cyclic Antidepressants

(4) When it comes to MAOIs, what two terms should spring to mind?

    (a) tyrosine and "wine and cheese effect"

    (b) tryptophan and "wine and cheese effect"

    (c) tyramine and "wine and cheese effect"

    (d) tyromean and the "wine and cheese effect"

(5) Emsam®/selegiline holds affinity for _____.

    (a) MAO-A

    (b) MAO-B

    (c) MAO-A & MAO-B

    (d) MOA-B

ANSWERS

(1) The correct answer is "a," glutamate is not a monoamine. It is an amino acid.

(2) The correct answer is "d," all the above. The driving influences behind changing MOA's that have been used for the last forty-plus years, is an attempt to make them more indicator driven (based on neurotransmitter action), reduce reliance on multiple medications for the same application, and to ultimately reduce nonsensical jargon and to move toward "clearer" terminology.

(3) The correct answer is "a," accidental. The first MAOIs created a "happy accident," a state of happiness in people suffering from consumption/tuberculosis.

(4) The correct answer is "c," tyramine and "wine and cheese effect." Here is the problem with psychopharmacology. So many of the words look alike. Tyramine acts as a sympathomimetic which, when increased, causes a substantial increase in blood pressure, heart rate, and if unabated, intracranial hemorrhage. It is "mean," and "tyromean" is a fake word. Or as test construction specialists call it, a "plausible distractor."

(5) The correct answer is "b," MAO-B – not MAO-A. Choice "d," is a tricky misspelling. It is easy to confuse MOA (Mechanism of Action) and MAO (Monoamine Oxidase).

Frankly, Class, Everyone Feels Pretty Sad

The SRIs

☺ ☺ ☺

Groundbreaking. Modern cures. Truly miraculous. The cure of the 80s. Bottled sunshine. No more depression. The antidepressant solution. Say "Goodbye to depression." Pills-a-go-go for the 80s. Kinder and gentler. Blockbusters. So new it has not had time to be improved. SAFE. The perfect pill. So much hype. So much hope. So much disappointment.

In 1987, Eli Lilly presented with great fanfare Prozac®/fluoxetine to the world. The first SRI which was zimelidine was first sold in 1982 and developed by Arvid Carlsson. He thought it was just another old antihistamine with alleged antidepressant qualities. Luvox®/fluvoxamine followed zimelidine (pulled from the market for serious adverse drug reactions) and then Prozac®/fluoxetine (both derived from the antihistamine diphenhydramine) arrived. Then along came the other SSRIs, or is it the SRIs? Which is the "more correct" terminology?

SRI vs SSRI? What is the difference? SRI stands for Serotonin Reuptake Inhibitor. Technically, Selective Serotonin Reuptake Inhibitors (SSRIs) are a "special" kind of SRI. SSRIs hold both (1) a reuptake status and (2) an affinity for a "specific" or selective receptor, hence the special "Double S" status. For instance, Prozac®/fluoxetine is truly a SSRI. In addition to being a reuptake inhibitor, it has an affinity, specifically, selectively for the 5-HT$_{2c}$ (serotonin 2-C) receptor, an affinity which imbues a special quality.

SSRI is more of a historical and marketing term rather than a label with significant clinical meaning. Many believe SSRI's creation was nothing more than a marketing trick to demonstrate a brand-new difference from the old, standby TCAs around since the 1960s. Others think it was a marketing ploy to reintroduce Schildkraut's "Biogenic Amine Hypothesis" (from the 1960s). All that said, it is a "fuzzy" distinction, but technically, SRI is the super-family name, and SSRI is a subset within that super family. To that point, SRI is used throughout this programmed-learning text; however, most people, including some very sophisticated textbooks, refer to the entire family as "SSRI" as opposed to the more accurate term, "SRI." Either class name one chooses to use is a no-penalty choice. Neither is preferred over the other.

SRI and SSRI distinction aside, this "nomenclature" is now out of date. With current standards and practices seeking to change the way mechanisms of action are written, focusing more on psychopharmacology rather than mental illness, the new "Neuroscience-based Nomenclature (NbN)" resolves some of this fuzziness by delineating all SSRIs/SRIs with one label, S-RI. The good news SRI (sans hyphen) and S-RI both translate to serotonin reuptake inhibitors. No more selective. While all SRIs are in some ways "the same," they are each in their own way different.

The following discussion focuses on the uniqueness of each SRI. Note the mnemonic ("Frankly, Class, Everyone Feels Pretty Sad"), this is not the order in which the medications were introduced into the market, but it is an easy way to remember all the SRIs. Some psychopharmacology instructors advise there is so little difference in the SRIs, that one could post them all on a dartboard and throw a dart to select the medication. No matter what one is attempting to treat – major depressive disorder, obsessive-compulsive disorder, premenstrual dysphoric disorder, anxiety, panic disorder, and social phobias among others – step up, throw a dart. This is neither best practice nor an effective way to prescribe and to maintain patient safety. While they all have the same mechanism of action and are used to treat the same litany of emotional concerns, each has its peculiarities and quirks.

The first entry on the dartboard, and the first to be introduced in the United States on December 29, 1987, Prozac®/fluoxetine, is the best place to start. It is the one that purportedly would change the world of mental health.

(1) Prozac®/fluoxetine. This was the first on the market and the first to go generic (2001). It has been around the longest and has been studied the most. Admittedly, it is not the best of the SRI antidepressants, but it had the benefit of prime marketing. To that point, Prozac®/fluoxetine has almost become prototypic or a general archetype for all antidepressants. When someone says, "You forgot to take your Prozac®." they mean, "Take an antidepressant!" Prozac®/fluoxetine has become the symbol for the illusory "perfect antidepressant." Probably the most redeeming quality for this celebrated medication is its long half-life. As a result, it is much easier to terminate this medication with minimal to no withdrawal effects. The flip side of the coin would indicate that Prozac®/fluoxetine is the only antidepressant

that requires a six- to eight-week period for a full therapeutic trial (most SRIs need a

four- to six-week trial).

SEROTONIN REUPTAKE INHIBITORS "Frankly, Class, Everyone Feels Pretty Sad" (Generic Mnemonic)		
Brand/Year Introduced in US	Generic	MOA[6]/NbN[7]
Prozac®/1987	fluoxetine	SSRI, selective serotonin reuptake inhibitor/S-RI, serotonin reuptake inhibitor
Celexa® /1998	citalopram	SRI, serotonin reuptake inhibitor/S-RI, serotonin reuptake inhibitor
Lexapro®/2002	escitalopram	SRI, serotonin reuptake inhibitor/S-RI, serotonin reuptake inhibitor
Luvox®/1994	fluvoxamine	SRI, serotonin reuptake inhibitor/S-RI, serotonin reuptake inhibitor
Paxil®/1992	paroxetine	SRI, serotonin reuptake inhibitor/S-RI, serotonin reuptake inhibitor
Zoloft®/1991	sertraline	SRI, serotonin reuptake inhibitor/S-RI, serotonin reuptake inhibitor

Prozac®/fluoxetine has a special affinity for the 5-HT_{2c} receptor, which

indirectly increases norepinephrine (and to some degree dopamine transmission).

Prozac®/fluoxetine is really a SNRI, serotonin norepinephrine reuptake inhibitor

(Ghaemi, 2019; Stahl, 2017).

---

[6] Mechanism of Action
[7] Neuroscience-based Nomenclature (NbN) from *Stahl's Essential Psychopharmacology Prescriber's Guide*, 6th Edition.

There is one other unique aspect to Prozac®/fluoxetine; it has substantial drug interactions via the CYP450 system. It is an enzymatic inhibitor when mixed with other medications.

Many of the side effects associated with Prozac®/fluoxetine are also side effects of the other SRIs. Here is an SRI side-effect mnemonic:

- **S** → in**S**omnia

- **S** → **S**exual (Lack of sexual desire)

- **R** → **R**estlessness (Mental and physical)/R_x interactions

- **I** → g**I** (gastrointestinal) distress

- **S** → **S**uicide

<u>(2 & 3) Celexa®/citalopram and Lexapro®/escitalopram</u>. These two drugs became available in the United States in 1998 and 2002, respectively. To understand these medications, start with the generic names – citalopram and escitalopram. What is the difference? Once again, it has to do with "S," but in this case, "es." Standby for a biochemistry lesson.

Approximately 50% of medications have a right and left side. Referred to as chirality (derived from Greek, "kheir" for "hand"), a property of asymmetry, an object is chiral if it is distinguishable from its mirror image. Right and left hands held together mirror each other; they are asymmetrical; one cannot superimpose one on the other. Drug molecules are the same way. They are "enantiomers," chiral molecules that are mirror images of each other. That means drugs can have, a left and right side. Another similar, yet slightly different word for chirality is stereoisomer.

The left side of the molecule is delineated "l", "levo," or "es" (for left, Latin for "sinister," meaning "malevolent, underhanded"). The right side of the molecule is delineated simply with "d" or "dextro" (from Latin, "dexter"). Look at the generic name of this drug, "d, l-amphetamine." This nomenclature advises that it is the racemic mixture and includes both the right ("d") and left side ("l") of the drug molecule. **Es**citalopram (note the underlined "es") is the left side only of the citalopram molecule. It does NOT include the right side. Celexa®/citalopram is both the left and right side of the molecule. Technically speaking, Lexapro®/escitalopram is an active enantiomer.

Generally (keyword, "generally"), it is the left side of the drug moiety that exerts the therapeutic effect; however, that does not hold in every case. There are exceptions. For amphetamines, it is the "d" or "dextro" side that is usually three times more beneficial psychotropically.

Imagine two drugs. One that is both sides of the drug molecule; and one that is the same drug, but only the left side. Which drug would present with the most side effects? Both sides/racemic mixture or just one therapeutic side of the drug (either "l/es/levo" or "d/dextro")? The drug with only the therapeutic side will exhibit less side effects. If it is a "both-sided," racemic mixture drug, one side will provide the therapeutic effect; the other side will guarantee additional side effects and possibly little to no therapeutic effect. This is exactly the case with "both-sided" Celexa®/citalopram and its half-brother, "one-sided" Lexapro®/**es**citalopram.

Celexa®/citalopram, a garden-variety SRI, is used to treat the same disorders as Lexapro®/escitalopram. However, Celexa®/citalopram has one big warning that demonstrates how the racemic mixture may cause more side effects than the left side, the more therapeutic side. Celexa®/citalopram has one warning from the FDA that Lexapro®/escitalopram does not have.

On August 24, 2011, the FDA issued the following warning for Celexa®/citalopram. The FDA warned prescribers that Celexa®/citalopram "should no longer be used at doses greater than 40 mg per day because it can cause abnormal changes in the electrical activity of the heart." Moreover, a dosage of 20 mg for elderly patients is the appropriate ceiling dosage of this medication. It is useful in the elderly, but the FDA admonition must be followed.

There is one other important point concerning Celexa®/citalopram. It is probably the truest "SRI" of the entire cadre of "SRIs." Whereas Prozac®/fluoxetine has some noradrenergic qualities, and Zoloft®/sertraline has some dopaminergic qualities, Celexa®/citalopram is the purest of the "SRIs," meaning it blocks the reuptake of serotonin almost exclusively (Ghaemi, 2019).

Often the question arises, "Which is the best of the SRIs?" Many problems with that query. First off, there is no "best." Each has abilities, and each has difficulties. Neither is a perfect pill. Some have peculiar side effects. Yet in a risk-benefit equation which is the best medication for the patient? Disclaimers aside, if a prescriber is looking for a drug with a longer half-life (at least 30 hours, for easier withdrawal), few CYP450 interactions, less sexual side effects, and limited GI distress, then the drug that represents the half side of the drug molecule,

Lexapro®/escitalopram, is the winner. This is especially true if the medication is being prescribed for a patient with existing heart conditions. Compared to its half-brother and the entire SRI family, Lexapro®/escitalopram is thought by many to be the antidepressant of choice (Ghaemi, 2019).

(4) Luvox®/fluvoxamine. This SRI is typically reserved for obsessive-compulsive diagnoses (OCD). It is potent and highly serotonergic with few other benefits. Although it is just another garden-variety SRI, it is a powerful CYP450 inhibitor. Translation: It does not interact well with numerous other medications.

(5) Paxil®/paroxetine. Clinicians have a love-hate relationship with this medication. It has a short half-life (around 15 to 21 hours), so it is challenging to withdraw from this medication. If the patient misses one dose, withdrawal effects follow by the second day. Paxil®/paroxetine has moderate anticholinergic effects (DUCTS: dry mouth, urinary retention, constipation, tachycardia, and sedation). Paxil®/paroxetine initially gained popularity because of its antianxiety (anxiolytic) effects.

(6) Zoloft®/sertraline. This medication seems to be a cardiac and obstetrics' favorite. For cardiologists, perhaps this medication is prescribed as a treatment for "heart-sick" depression, but more likely for SRI's antiplatelet effects. For obstetricians, it is the medication of choice for breastfeeding mothers, perhaps because it is the least bioavailable (approximately 40% — 45%). There is also some dopaminergic action associated with Zoloft®/sertraline. As a result, it is classified as a "SRI," but more likely it is a "SDRI." However, the new "NbN" terminology does not acknowledge this. Finally, Zoloft®/sertraline also acts on the "sigma receptors."

While these receptors' complete function is not clearly understood, it is thought that through this receptor Zoloft®/sertraline exerts an antianxiety (anxiolytic) effect.

Demonstrating the value of the new "NbN" nomenclature, two newer medications are "neither fish nor fowl." They are SRIs and more. These medications are Viibryd®/vilazodone and Trintellix®/vortioxetine. Now the mnemonic changes to read: "Frankly, Class, Everyone Feels Pretty Sad, Very Vague."

(7) Viibryd®/vilazodone. It is a not-so-distant cousin of the well-known drug Desyrel®/trazodone (an antidepressant, but typically used more off-label for sleep; note both drugs end in the suffix, "o-done"). It is a SRI with two mechanisms of action (hence the "marketing trick" using double letter "i's" in the spelling of Viibryd®). It is a RI, reuptake inhibitor; plus it is a partial agonist at the 5-HT$_{1A}$ receptor. Because it is double acting, it is referred to as a serotonin partial agonist reuptake inhibitor (SPARI). It functions exactly the same as other SRIs, but with a little more "oomph." By partially agonizing the 5-HT$_{1A}$ receptor, it boosts serotonin an additional 20%. Interestingly, the 5-HT$_{1A}$ receptor is thought to be the one receptor out of the total fourteen 5-HT receptors that exerts the antidepressant effect. If one wanted to make it really complicated, this drug's nomenclature could be described as a "SPASSRI." There is no such naming. This makes the need for a more standard nomenclature obvious. As a result, the "NbN" terminology for this drug's mechanism of action is S-MM, serotonin multimodal (Ghaemi, 2019).

(8) Trintellix®/vortioxetine. In 2013, this medication was originally called "Brintellix," but the name was changed in 2016 to Trintellix®. The name change was

because the original name, Brintellix®, was too similar to AstraZeneca's

Brilinta®/ticagrelor, an antiplatelet medication.

SEROTONIN REUPTAKE INHIBITORS PLUS TWO "Frankly, Class, Everyone Feels Pretty Sad, Very Vague."		
Brand/Year Introduced in US	Generic	MOA[8]/NbN[9]
Prozac®/1988	fluoxetine	SSRI, selective serotonin reuptake inhibitor/ S-RI, serotonin reuptake inhibitor
Celexa® /1998	citalopram	SRI, serotonin reuptake inhibitor/ S-RI, serotonin reuptake inhibitor
Lexapro®/2002	escitalopram	SRI, serotonin reuptake inhibitor/ S-RI, serotonin reuptake inhibitor
Luvox®/1994	fluvoxamine	SRI, serotonin reuptake inhibitor/ S-RI, serotonin reuptake inhibitor
Paxil®/1992	paroxetine	SRI, serotonin reuptake inhibitor/ S-RI, serotonin reuptake inhibitor
Zoloft®/1991	sertraline	SRI, serotonin reuptake inhibitor/ S-RI, serotonin reuptake inhibitor
Viibryd®[3]/2011	vilazodone	SPARI, serotonin partial agonist reuptake Inhibitor/ S-MM, serotonin multimodal
Trintellix®[3]/2016	vortioxetine	SSRI+[10]/ S-MM, serotonin multimodal

[8] Mechanism of Action
[9] Neuroscience-based Nomenclature (NbN) from *Stahl's Essential Psychopharmacology Prescriber's Guide*, 6th Edition.
[10] Old nomenclature is less accurate; clearer delineation using new nomenclature needed and explained in upcoming modules.

Like Desyrel®/trazodone and Viibryd®/vilazodone, Trintellix®/vortioxetine is an SRI, plus more – once again making it difficult to nomenclate. It is a full agonist on the 5-HT_{1A} receptor. Additionally, it blocks the 5-HT_3 and 5-HT_7 receptors. The 5-HT_3 receptor is present in abundance in the gastrointestinal system and regulates gut motility, secretions, and peristalsis (involuntary constriction/relaxation of intestinal muscles). The 5-HT_7 receptor affects thermoregulation, circadian rhythms, learning, memory, and sleep. It is further conjectured that this receptor might possibly be important in the treatment of depression. Because this medication works on numerous 5-HT receptors, "NbN" has identified this medication as "Serotonin Multimodal (MM)." However, the exact clinical relevance of these receptors to the diagnosis is unclear.

One important point about this medication concerns its half-life. Trintellix®/vortioxetine has an exceptionally long half-life (66 hours) which could minimize withdrawal. While this medication is touted as unique, it is not that unique. Qualities of this medication can be found in Desyrel®/trazadone, Viibryd®/vilazodone, and even Remeron®/mirtazapine (Ghaemi, 2019). More information on these "atypical antidepressants" in Module 41.

Overall, the SRI family is safe with a wide therapeutic index. Expect difficult withdrawal, for all excepting Prozac®/fluoxetine which has a very long half-life. Numerous side effects relate to mood, appetite, pain, panic, suicide, sleep, intra-ocular pressure, and fractures. Sexual problems occur 80% of the time. More similarities than differences, may be more placebo than substance, and combined with therapy for moderate to severe depression, they are a good start.

QUIZ TIME

MATCHING:

(1) _____ Prozac®/fluoxetine

(2) _____ Zoloft®/sertraline

(3) _____ Paxil®/paroxetine

(4) _____ Celexa®/citalopram

(5) _____ Lexapro®/escitalopram

(6) _____ Luvox®/fluvoxamine

(7) _____ Viibryd®/vilazodone

(8) _____ Trintellix®/vortioxetine

A. Left sided

B. Cousin of trazodone

C. OCD

D. 5-HT$_3$ and 5-HT$_7$ receptors

E. First in US

F. The purest SRI

G. DUCTS

H. Favored by obstetricians

(9) True or False? The shorter the half-life, the more difficult it is to withdraw from the antidepressant.

(10) True or False? If the patient has mild to moderate depression, drug treatment only is recommended.

ANSWERS

(1) Prozac®/fluoxetine, "E," first in United States; (2) Zoloft®/sertraline, "H," favored by obstetricians; (3) Paxil®/paroxetine, "G," DUCTS, most anticholinergic; (4) Celexa®/citalopram, "F," The purest SRI; (5) Lexapro®/escitalopram, "A," Left sided; (6) Luvox®/fluvoxamine, "C," preferred treatment for OCD; (7) Viibryd®/vilazodone, "B," Cousin of Desyrel®/trazodone; (8) Trintellix®/vortioxetine, "D," 5-HT$_3$ and 5-HT$_7$ receptors, selectively blocks.

(9) TRUE. The shorter the half-life, the more difficult it is to withdraw from the medication.

(10) FALSE. For mild to moderate depression, the treatment of choice is not medication only, therapy is considered first line. If the depression (as indicated on a depression measurement scale) is moderate to severe, then the treatment of choice is therapy plus medication. There is no situation in the research literature where drug treatment only is the treatment of choice for depression (Fournier, et al., 2010).

A Balancing Act

Discontinuation and Serotonin Syndromes

☺ ☺ ☺ ☺ ☺

It is all about the body's balance. Homeostasis. Manipulate the body's balancing act, and pronounced symptoms will appear. Once again, a teeter-totter, for every up, there is a possible down below. Too little, the body goes into withdrawal. Too much, the body advances too rapidly, and a new set of stormy indicators emerge. Stop taking serotonergic medications abruptly, and the dramatic reduction causes discontinuation syndrome or withdrawal. Take too much, the other side of the coin, and serotonin storms the system and poisons the body. Serotonin withdrawal is nasty; serotonin poisoning is a medical emergency. As a rule, all antidepressants should be tapered slowly, NEVER stopped abruptly.

The exception is Prozac®/fluoxetine which can be stopped as quickly as it was started. That said, some sources continue to suggest that it, too, should be tapered. So, when the patient asks, "Can I stop taking Prozac®?" The answer is a judicious, "Yes," with appropriate caveats based on the patient's needs. Because it has a long half-life, it takes the body a long period to taper, and slowly the body returns to original homeostasis. The result: Few to no discontinuation symptoms (Stahl, 2017).

Discontinuation/withdrawal signs and symptoms include:

- Dizziness
- Anxiety
- Muscle aches
- Myoclonus
- Nervousness
- Insomnia
- Fever
- Hypertension
- Nausea
- Tachycardia

The reason for many of these symptoms (including insomnia, fever, hypertension, and myoclonus) is hypothalamic dysregulation due to abrupt withdrawal. The hypothalamus is a small (the size of a sugar cube; for reference, its partner, the pituitary, the master gland, is the size of a pea), critical area of the brain and the seat of regulatory functions including temperature, weight, emotions, sleep cycle, sex drive, and more.

Often the patient misunderstands their withdrawal or discontinuation symptoms. To explain, if the antidepressant grants depression relief, withdrawal symptoms will feel like a rapid return of depression. The patient unwittingly assumes that the emerging symptoms, feelings of horrible depression, are a signal that they really do need their medication. In other words, "I stopped taking my medication; my depression returned; I must need these medications." They immediately resume the antidepressant, and the withdrawal symptoms go away. This becomes "proof" they are depressed and require the medication.

Best advice here is never prescribe a medication without honoring two caveats. First, never prescribe a pill without having a plan in mind to help a patient withdraw from the drug. Second, explain that if they stop the medication on their own, the symptoms that emerge result from withdrawal and not advanced depression.

Two little tricks to watch out for. If a patient returns for a medication check and is crying, weeping, and/or very tearful, casually ask: "Are you taking your antidepressants (SRI) as prescribed?" Do not be surprised if they say, "No." This weepiness is a clue that they are trying to stop on their own without the prescriber's help. If the patient is taking a Tri-Cyclic Antidepressant (TCA), comes in for a med check and complains of sudden stuffy nose and congestion (not a sore throat, that may be a sign of "agranulocytosis" typically from other more complicated medications, or it could be influenza or a cold), just ask, "Are you still taking your TCA?" Do not be surprised when the client says, "No." This congestion is the result of cholinergic rebound from abruptly stopping a Tri-Cyclic Antidepressant (TCA).

At the other extreme, what happens when one takes too many serotonergic medications? Imagine an elderly lady is bereft over the death of her only son. Her husband died some years earlier, and she has been taking Valium®/diazepam daily ever since. Her prescriber suggests she take a "SRI" to take the edge off her grieving. She agrees. She expects this new SRI to work like her Valium®/diazepam within twenty minutes (recall that Valium®/diazepam is a psychoactive). SRIs do not work that quickly. She takes her first SRI and given her expectation, it does not work within twenty minutes. She continues to take them, too many, too fast. Serotonin builds up in her system, and she expresses signs and symptoms of "Serotonin Syndrome" or "Serotonin Storm." It is now a medical emergency.

Another common, fictional example, but unfortunately, one that plays out every day. A wounded warrior suffering from Post-Traumatic Stress Injury (PTSI; a term preferred to PTSD) has been prescribed Valium®/diazepam. It grants relief in twenty to thirty minutes working directly on the central nervous system. To eventually move this soldier from this "problem drug" to a more benign one, the prescriber adds a SRI to his PTSI treatment regimen. The warrior takes the first SRI; it does not work as expected. It does not grant relief in twenty minutes. For relief, another is taken, and another, and another, and another. The result, too much serotonin builds up in the system; the dose far exceeds the body's capacity to break it down; the cure, as Paracelsus suggested, moves from therapeutic to toxic dose, and serotonin syndrome emerges. It is now an emergency and requires medical treatment.

The signs and symptoms of 5-HT Syndrome, Serotonin Syndrome, Serotonin Storm (Acute Crisis) are as follows (bold for emphasis; explanation follows):

- **Hyperreflexia** (overactive/over-responsive reflexes, twitching or spastic tendencies)
- **Agitation**
- **Diaphoresis** (excessive sweating)
- **Tremor**
- Mental status changes
- Shivering
- Myoclonus (involuntary muscle contractions)
- Explosive diarrhea
- Poor coordination
- Fever

While there are no lab tests for 5-HT, Serotonin Syndrome, it is a medical emergency. From this list, the most reliable indicators of 5-HT Syndrome are the first four (bolded in list): (1) hyperreflexia, (2) agitation, (3) diaphoresis, and (4) tremor (Sternbach, 1991). If all four are present, 5-HT syndrome must be ruled out. Extreme cases require medical attention and likely hospitalization. This syndrome can result from taking too much of a serotonin medication too quickly, mixing serotonin medications with other serotonin medications, mixing serotonin medications with monoamine oxidase inhibitors, and overdose.

Serotonin syndrome is unmistakable. Once a clinician sees it, they will never forget it. Take Sternbach's top four symptoms and add to it "explosive diarrhea," and the sensory input from this condition renders it unforgettable.

Withdrawal or discontinuation syndrome unwittingly happens more often than serotonin syndrome. Withdrawal is usually misinterpreted as "I need this medication. See how depressed I get when I don't take it." To alleviate the withdrawal symptoms, the patient resumes the medication. Unfortunately, numerous prescribers do not realize that the patient's attempt to stop the medication can result in withdrawal symptoms. When the patient explains how depressed they became when they stopped the medication, numerous prescribers do not recognize heightened depression as withdrawal symptoms. Instead, they increase the dose.

QUIZ TIME

(1) True or False? A stopped-up nose could signal a patient's attempt to stop taking a Tri-Cyclic Antidepressant (TCA) on their own without guidance.

(2) True or False? Weepiness could signal a patient's attempt to stop taking a SRI on their own without guidance.

(3) 5-HT Syndrome is also called _____.

    (a) serotonin squall

    (b) serotonin attack

    (c) serotonin storm

    (d) serotonin breakthrough

(4) Which of the following is/are indicator(s) of 5-HT Syndrome?

    (a) hyperreflexia

    (b) agitation

    (c) diaphoresis

    (d) tremor

    (e) all the above, plus bonus: explosive diarrhea.

(5) True or False? Often withdrawal symptoms are misinterpreted by the patient as the need for their antidepressant or the need for an increased dose.

ANSWERS

(1) TRUE. This is called cholinergic rebound. A sudden stuffy nose could signal the patient's desire to stop taking a TCA on their own without prescriber guidance.

(2) TRUE. Weepiness is a classic example that the patient is attempting to stop a SRI on their own, without prescriber guidance.

(3) The correct answer is "c," serotonin storm. It is a medical emergency and requires treatment immediately.

(4) The correct answer is "e," all the above. In fact, at the risk of indelicacy, the nickname for 5-HT syndrome is "scheisse & shivers." "Scheisse" is German for fecal matter. Not that it sounds better, it just looks better in print – but there is nothing attractive about explosive diarrhea resulting from too much serotonin.

(5) TRUE. If a patient is not aware of discontinuation or withdrawal syndrome's symptoms, they have no frame of reference to interpret their increased depression. The logical fallacy to get rid of withdrawal symptoms is to take more antidepressant. The result, "I stopped my meds. Look how bad I feel. I must really need my antidepressant."

Module 39

Strange and True

Abstruse SRI Side Effects

😊 😊 😊

Remember the "SRI" side-effects mnemonic from an earlier module – in**S**omnia, **S**exual, **R**estlessness, **R**x Interactions, gastro**I**ntestinal distress, and **S**uicide? There are four more side effects — rarely taught, abstruse, obscure, but more common than expected, strange ones, and true. As a result, few clinicians would even know that they are related to SRI use. If there is serotonin in the drug's mechanism of action (or even hiding somewhere in the formulary), be on the lookout for these four curious side effects: (1) Amotivational syndrome, (2) Syndrome of Inappropriate Antidiuretic Hormone (SIADH; hyponatremia), (3) jaw pain, and (4) fractures.

(1) <u>Amotivational Syndrome or Serotonin-Induced Indifference</u>. Remember, the SRIs are poorly named. They are not antidepressants; at best, they are "serenics" (from the root word, "serene"). Imagine a patient diagnosed with moderate-to-severe depression has been placed on a garden-variety antidepressant, chock full of serotonin. Two weeks later, the patient returns for a drug check. They are asked the ubiquitous question, "How are you feeling?" They commonly answer usually something like, "I feel nothing" or "I feel numb." Other reports include phrases like, "Why can't I cry?" and "I feel like my depression is trapped." The depression has not resolved;

rather, any emotional richness is blocked. The reason is found in the relationship between serotonin and dopamine.

There is a homeostatic balancing act between serotonin and dopamine. In the patient's case, when the medication indirectly agonizes and increases serotonin (via reuptake blockade), serotonin is stimulated; the balance is upset; and the body compensates by decreasing dopamine. When the antidepressant increases serotonin, dopamine is blocked, and the result is a "numbness," a "nothingness," a "serene calmness," or to put in technical jargon, "amotivational syndrome." To resolve this problem, many clinicians advise this is nothing more than an activating aspect of the medication and will go away when the receptors down-regulate. Perhaps it is also an indication that the dose is too high (Sansone & Sansone, 2010).

(2) <u>Syndrome of Inappropriate Antidiuretic Hormone or Hyponatremia</u>. Syndrome of Inappropriate Antidiuretic Hormone (SIADH) or hyponatremia ("hypo" means "reduced," and "natrium" is Latin for sodium; "emia" is Greek for "blood" = reduced sodium in blood) is brought about by an excess of water rather than a deficit of sodium. This condition, SIADH, causes the body to make too much antidiuretic hormone (ADH), a chemical that regulates the body's water balance. Too much ADH causes the body to retain water. Too much water volume tends to dilute the amount of other chemicals/minerals in the blood, including salt. This imbalance makes it difficult for organs to function properly. SIADH is caused by numerous conditions like cancer, spinal cord injuries, chronic obstructive pulmonary disease (COPD), family history, too much pain or stress on the body, and antidepressants. SIADH can be diagnosed via blood and urine tests. Hyponatremia has been associated in case

reports, observational studies, and case-controlled studies associating SIADH with SSRI use with an incidence of 0.5% to 32%. While it may be related to paracrine signaling, the cause or relationship between SRIs and SIADH is not clearly understood (El-Merahbi, et al., 2015). The key phrase to remember is "older adults," and hyponatremia is another variable to monitor with this population (Jacob & Spinler, 2006).

(3) <u>Jaw Pain</u>. To understand this particular side effect, take a lesson from 3,4-methylenedioxymethamphetamine (MDMA), better known as Ecstasy. If one were able to attend a rave to watch the kids dancing, it is immediately apparent that many of them are sucking on pacifiers. There are two important reasons for this quasi-Freudian behavior. First, Ecstasy (MDMA) is sold coated on the nipple of the pacifier — an oral, enteral delivery system. Second, one main side effect of Ecstasy is an involuntary jaw clamping and biting one's cheek. The pacifier protects the abuser from this involuntary painful clamping, biting cheek. The explanation is found once again in the homeostatic relationship between serotonin and dopamine.

When serotonin increases, dopamine decreases. When dopamine is manipulated (either agonized or antagonized), it signals the nigrostriatal dopaminergic pathway. Ecstasy is highly serotonergic; therefore, dopamine will be significantly blocked. This blockade manifests as a movement disorder in the jaw, notably an involuntary clamping down. While the SRIs are not as potently serotonergic (like Ecstasy), their movement disorders are related to the jaw including biting cheek, bruxism (teeth grinding), even tooth loosening (tooth mobility). Teeth grinding and jaw clamping are counterintuitive side effects of SRIs; thus, a patient rarely brings it up to the mental health professional and is more likely to discuss it with their dentist. Nevertheless, it is a fair question ("Have you been grinding your teeth or biting your jaw?") to ask patients to whom an SRI has been prescribed (Wise, 2001).

(4) Fractures. It is always surprising to prescribers that such a thought-to-be-benign medication can cause bone fractures – especially in the elderly. Put another way, antidepressants have negative effects on bone, particularly bone mineral density and fracture risk. Additionally, these risks are enhanced by more prolonged exposure to SRIs. Current studies available do not clearly delineate the magnitude of the relationship between SRIs and negative bone effects, but all indicate some risk (Sansone & Sansone, 2012). Serotonin receptors have been found on osteoclasts, osteoblasts, and osteocyte cell lines suggesting that 5-HT has a regulatory effect on bone. While there are no formal recommendations regarding SRIs in risk populations – especially the elderly and those with osteoporosis histories and osteoporotic fractures – this is a concern that prescribers must monitor.

QUIZ TIME

(1) A patient just started taking a SRI and complains, "My emotions feel trapped." Chances are this patient is experiencing _____.

    (a) flat affect

    (b) amotivational syndrome

    (c) anaclitic depression

    (d) breakthrough treatment

(2) True or False? Extended use of SSRIs is associated with increased possibility of bone fractures.

(3) While doing the laundry, a pacifier fell from your college-aged son's favorite pair of jeans. It could be nothing, but just in case it is not, what thoughts should cross your mind?

    (a) Regressing toward a Freudian oral fixation

    (b) Sexual experimentation

    (c) Ecstasy

    (d) Leftover from babysitting his baby brother

(4) Hyponatremia means _____.

    (a) too much sodium to body fluid volume

    (b) too little sodium to body fluid volume

    (c) homeostatic level of sodium to body fluid volume

    (d) too much potassium to body fluid volume

(5) Which of the following is NOT a common side effect of SRIs?

    (a) Increased libido

    (b) R_x Interactions

    (c) Gastro-Intestinal Distress

    (d) Insomnia

ANSWERS

(1) The correct answer is "b," amotivational syndrome. Too much serotonin reduces normal dopamine levels. Reduction of dopamine creates a feeling of "numbness" or "trapped emotions."

(2) TRUE. Serotonin receptors have been found on osteoclasts, osteoblasts, and osteocyte cell lines suggesting that 5-HT has a regulatory effect on bone.

(3) The correct answer is "a," "b," "c," "d," or any combination. However, rule out Ecstasy.

(4) The correct answer is "b," too little sodium to body fluid volume. Translation: "Hypo" means reduced; "natrium" is Latin for sodium; "emia" is Greek for "blood." Putting it all together, hyponatremia means reduced sodium in the blood volume.

(5) The correct answer is "a," increased libido. SSRIs do NOT increase libido but decrease sexual interest. Hard to believe, but they anesthetize one's genitals.

Module 40

Something Old, Something New, Something Borrowed, Something Blue

TCAs and SNRIs

☺ ☺ ☺

In the 1950s, known as the "The Golden Age of Psychopharmacology," Thorazine®/chlorpromazine emerged as a godsend. Australian physician John Cade rediscovered lithium's therapeutic effects, Monoamine Oxidase Inhibitors (MAOIs) arrived for tuberculosis patients, and then the early 1960s heralded something brand new – "the antidepressant." Originally called a "thymoleptic" ("thy" from Greek "thymos," meaning feelings and passions; and "litic" from Greek, "lepsis," meaning "seizing"), the researchers had no clear idea what they had created. They hoped for another antipsychotic, something that would "seize or block" unwanted feelings and passions. When it did not work for psychoses, they were nonplussed (Healy, 2002).

Something Old. In March 1949, enter Alan Broadhurst, a new chemist for Geigy in Rhodes, England, who aspired to create a new, different, and unique drug. Playing around with the structure of Thorazine®/chlorpromazine, led him quite accidentally (after forty-two tries) to the creation of a new drug, labeled G22150, the least toxic of his many formulations. Its structure was iminodibenzyl (dibenzazepine) based – two benzene rings fused to an azepine group, giving it the "look" of three circles. He was manipulating a dye and discovered a "new" drug. All that remained was to find the disease it treated.

This selfsame logic exists today. For example, a chemical compound that blocked the histamine, H_3 autoreceptor, which controls gastric acid secretions, was discovered in 1983; then, voila, "Gastroesophageal Reflux Disorder" (GERD), a new disease contiguous with a marketable drug surfaced.

Broadhurst's first clinical trial of G22150 was on himself. Thinking it might be a sleeping pill, he arranged a small drug test with a local physician. The trial failed; it was not a sleeping pill. Undaunted he surmised, given the similar three-ring structure of Thorazine®/chlorpromazine, perhaps it could be a "neuroleptic" ("nerve seizing," the original name for antipsychotic medications; this terminology never caught on in the United States) medication like Thorazine®/chlorpromazine. It failed, but it gave the psychotic patients energy. According to accounts, they "skipped and sang" and were "oddly happy" (Healy, 2002). Ultimately, given the drug's three-ring structure, this "oddly happy" pharmaceutical accident would be known as "Tri-Cyclics" and eventually "Tri-Cyclic Antidepressants (TCAs)." It is the only antidepressant identified by its chemical structure. If it arrived on today's drug market, it would be a SNRI – serotonin norepinephrine reuptake inhibitor.

In 1963, G22150 became Tofranil®/imipramine, and "thymoleptic" eventually gave way to "antidepressant" after pharmaceutical researchers found a market for Tofranil®/imipramine. Even though Max Lurie coined the term antidepressant in 1952, it did not catch on. The prevailing view from the early 1950s until the early 1960s, including Broadhurst's opinion, was that the only appropriate way to treat "hospital depression" was in a hospital. Treating "community depression" locally, outside of a hospital, was unheard of and unthinkable; this approach did not exist until Tofranil®/imipramine spurred the idea.

However, the researchers at the time had no idea what they had created. This was long before randomized clinical trials were required prior to their arriving for public consumption. While it seems as if this class of medication has been around for ages, especially since they are used so regularly today, the whole class of Tri-Cyclic Antidepressants (TCAs) may feel modern but are 1960s old. All TCAs are similar to each other; however, each has individual idiosyncrasies, potencies, side effects, and specialties.

If a drug ends in "mine" (pronounced "mean"), "line" (pronounced "lean"), or "pin," the odds are good that the medication is a TCA. The entire class of TCAs is respectfully known as "old, dirty, and cheap." Or to put it more tactfully, TCAs are "time-tested, pharmacologically rich, and inexpensive." They are seven pharmacologically rich pills in one dose and have lots of "free extras," like weight gain, orthostatic hypotension (dizziness upon standing), and anticholinergic effects (DUCTS – dry mouth, urinary retention, constipation, tachycardia, sedation). They are also *not* recommended for patients with existing heart problems. They can affect the heartbeat's QTc interval, electrical conductivity, as seen on heart tracings. Because they can block ion channels on the heart, they have a narrow therapeutic index and can be potentially lethal in overdose. If a suicidal patient is given a two-week supply of TCAs, the prescriber has unwittingly placed the patient at risk. Once upon a time, they were inexpensive. However, because TCAs, especially Elavil®/amitriptyline, have been rediscovered and prescribed regularly today, the price has increased dramatically.

While all antidepressants are equally efficacious for depression (FDA approved for depression and endogenous depression), many of them are also used off-label for fibromyalgia, headache, lower back pain, anxiety, OCD, itching, insomnia, treatment resistant depression, and others. These medications are seen regularly as a treatment for pain. Pain relief, usually about 30%, is likely the result of "their impact on blocking sodium and calcium channels peripherally, not their effects on norepinephrine and serotonin" (Ghaemi, 2019, p. 4).

Additionally, several of the TCAs have idiosyncratic specialties:

- Surmontil®/trimipramine, Elavil®/amitriptyline, Sinequan®/Silenor®/doxepin are the _most_ sedating.
- Tofranil®/imipramine is indicated for enuresis (bedwetting).
- Anafranil®/clomipramine is the most serotonergic and indicated for obsessive-compulsive disorder.
- Elavil®/amitriptyline and Tofranil®/imipramine are also sedating along with weight gain and orthostatic hypotension — not a good choice for the elderly.
- All TCAs (actually most antidepressants) decrease seizure threshold – making it easier for patients to seize.

Something Borrowed. Norpramin®/desipramine and Pamelor®/nortriptyline were chemically "borrowed" from older TCAs. For years, they have been "dumped in" with all the TCAs, but they are different. First, look at their generic names, desipramine and nortriptyline, for a clue as to its original TCA namesake. Desipramine is the active metabolite of imipramine, and nortriptyline is the active metabolite of amitriptyline. Using the prefix "nor" and "des" implies that they are "structural analogs" (the desmethyl metabolite) of the original TCA. The active metabolite is

borrowed from the guarantor, the original TCA.

TRI-CYCLIC ANTIDEPRESSANTS (TCAs)		
Brand (Year Approved in US)	Generic	MOA[11]/NbN[12]
Surmontil® (1979)	trimipramine[13]	SNRI, serotonin norepinephrine reuptake inhibitor/DS-Ran, dopamine serotonin receptor antagonist
Elavil® (1961)	Amitriptyline[13]	SNRI, serotonin norepinephrine reuptake inhibitor/SN-MM, serotonin, norepinephrine multimodal
Anafranil® (1990)	Clomipramine[13]	SNRI, serotonin norepinephrine reuptake inhibitor/S-RI, serotonin reuptake inhibitor; the NbN-2 app identifies it as a serotonin dopamine antagonist[15]
Norpramin® (1964)	desipramine[14]	SNRI, serotonin norepinephrine reuptake inhibitor/N-RI norepinephrine reuptake inhibitor
Sinequan® (1969)	Doxepin[13]	SNRI, serotonin norepinephrine reuptake inhibitor/SN-MM, serotonin norepinephrine multimodal
Tofranil® (1959)	Imipramine[13]	SNRI, serotonin norepinephrine reuptake inhibitor/SN-RI, serotonin norepinephrine reuptake inhibitor
Pamelor® (1964)	Nortriptyline[14]	SNRI, serotonin norepinephrine reuptake inhibitor/SN-RI, Serotonin norepinephrine reuptake inhibitor; the NbN-2 app identifies it as a Norepinephrine Reuptake Inhibitor, N-RI[15]
Vivactil® (1964)	Protriptyline[14]	SNRI, serotonin norepinephrine reuptake inhibitor/SN-RI, serotonin norepinephrine reuptake inhibitor

---

[11] Mechanism of Action (MOA)/not to be confused with Monoamine Oxidase (MAO)

[12] Neuroscience-based Nomenclature from *Stahl's Essential Psychopharmacology Prescriber's Guide*, 6th Edition.

[13] Tertiary amine

[14] Secondary amine

[15] Per NbN-2 App

Norpramin®/desipramine and Pamelor®/nortriptyline are neither TCAs nor SNRIs (recall if TCAs arrived at market today, they would be classified as SNRIs). These two medications are NRIs – norepinephrine reuptake inhibitors, even though old classifications refer to them as "SNRIs." There are two main classes of TCAs: (1) tertiary amines, which prefer serotonin and (2) secondary amines which are more selective for norepinephrine. Norpramin®/desipramine and Pamelor®/nortriptyline are secondary amines. These two secondary amines, "borrowed" from their original guarantors, offers three benefits. First, they have the least anticholinergic side effects (DUCTS – dry mouth, urinary retention, constipation, tachycardia, sedation). Second, they cause less confusion. Third, they decrease the probability of orthostatic hypotension (Ramey, et al., 2014).

Something New (or Something Newer). Older TCA's made their first appearance in the late 1950s, early 1960s. Newer serotonin norepinephrine reuptake inhibitors, arrived 30+ years later. While they do not have a three-ring configuration, they do share the same mechanism of action.

Effexor®/venlafaxine, appeared in the United States in 1993 and is FDA approved to treat depression (more efficacious for moderate to severe cases), generalized anxiety disorder, social anxiety disorder (social phobia), and panic disorder. Effexor®/venlafaxine is a SRI (at low doses) masquerading as a SNRI. To that point, the dosage of Effexor®/venlafaxine makes it three pills in one. At a lower dose (less than 150 mg), it is predominantly a serotonin reuptake blocker; at moderate doses (150 mg – 300 mg), it is a serotonin norepinephrine reuptake blocker; and at higher doses (greater than 300 mg) it is a serotonin norepinephrine dopamine

282

reuptake blocker. Above 450 mg, seizures become an adverse drug reaction. Effexor®/venlafaxine is 30 times more potent for serotonin reuptake blockade versus norepinephrine reuptake blockade (Ghaemi, 2019). Given their relatively short half-life, discontinuation is difficult, painful, and severe. Many clinicians agree that Effexor®/venlafaxine may be among the worst agents for causing withdrawal syndrome, even with the XR (extended release) formulation (Ghaemi, 2019).

Concerning bipolar depression, Effexor®/venlafaxine is at least twice as likely to "flip" or cause mania than other SRIs. To that point, and demonstrated by randomized trials, "this medication should not be prescribed at all in bipolar depression" (Ghaemi, 2019, p. 204).

However, there is a bigger problem associated with this medication: hypertension. Effexor®/venlafaxine "raises the blood pressure and has cardiovascular risks of sudden cardiac death" (Ghaemi, 2019, p. 95). While this blood pressure increase has been downplayed by the manufacturer (claiming a mean increase of only up to 3 mmHg), given the number of other choices, it is easy to avoid this medication with hypertensive patients.

Next, look at Effexor's® generic name, venlafaxine; now look at the generic name of the second member of the SNRI family, Pristiq®/desvenlafaxine. Venlafaxine vs desvenlafaxine. Effexor®/venlafaxine is the parent of Pristiq®/desvenlafaxine; venlafaxine is metabolized to O-desmethylvenlafaxine or desvenlafaxine. This active metabolite, Pristiq®/desvenlafaxine, was approved as an antidepressant in 2008. This family relationship makes these two drugs similar in efficacy, pharmacodynamics, and side effect profile. However, its pharmacokinetics ("ADME") are different (Liebowitz & Tourian, 2010). They are both SNRIs, and both are FDA approved for the treatment of major depressive disorder. Withdrawal from both can be difficult to severe. Risk and benefits of Pristiq®/desvenlafaxine are not noticeably different from Effexor®/venlafaxine.

Then there was the "bladder stabilizer" that failed to sell. What to do? Market it as an antidepressant. Cymbalta®/duloxetine entered the US market in 2004. Even though it has been referred to as "souped-up Prozac," it is more of a true SNRI. It is nine-fold more potent for serotonin reuptake blockade than norepinephrine reuptake blockade. It is FDA approved to treat major depressive disorder, diabetic peripheral neuropathic pain (DPNP), fibromyalgia, generalized anxiety disorder (acute and maintenance), and chronic musculoskeletal pain. The good news is that it is shown to reduce cognitive symptoms in the elderly and is less likely to increase blood pressure than venlafaxine. The bad news is that it is a CYP450 2D6 inhibitor. That means taking Cymbalta®/duloxetine can cause blood levels of TCAs and several antipsychotics to increase. Like the other SNRI family members, patients who discontinue this medication can experience severe withdrawal symptoms.

The new kid on the block, Fetzima®/levomilnacipran, entered the US market in 2013 and is another SNRI, emphasizing "N," norepinephrine. It is FDA approved to treat major depressive disorder. To better understand this medication, here is a chance to use biochemistry knowledge from a previous module.

Remember that approximately 50% of medications have a left and right side? Recall that the left side is abbreviated as "l," "es," and "levo." This applies to Fetzima®/levomilnacipran. Look at the generic name "levo" and "milnacipran." This medication is the "left," "levo" side of the milnacipran molecule. Milnacipran is the racemic mixture – both left and right sides. Fetzima®/levomilnacipran is the left side of the medication. This left enantiomer of Savella®/milnacipran has been used in Europe for years as an efficacious treatment for depression. Because Savella®/milnacipran is more potent for norepinephrine, high blood pressure is a prescribing concern. Problems aside, it is a strong choice for patients who are experiencing depression associated with chronic pain.

Many consider Fetzima®/levomilnacipran as a "Me, too." drug. This is a drug that is usually structurally similar to other drugs on the market. There may be negative connotations, but "Me, too." drugs drive prices down. To that point, Fetzima®/levomilnacipran has not been proven more effective than other SNRIs.

SEROTONIN NOREPINEPHRINE REUPTAKE INHIBITORS (SNRI)

Brand (Year Approved in US)	Generic	MOA[16]/NbN[17]
Effexor® (1993)	venlafaxine	SNRI, serotonin norepinephrine reuptake inhibitor/SN-RI, serotonin norepinephrine reuptake inhibitor
Pristiq® (2008)	desvenlafaxine	SNRI, serotonin norepinephrine reuptake inhibitor/SN-RI, serotonin norepinephrine reuptake inhibitor
Cymbalta® (2004)	duloxetine	SNRI, serotonin norepinephrine reuptake inhibitor/SN-RI, serotonin norepinephrine reuptake inhibitor
Fetzima (2013)	levomilnacipran	SNRI, serotonin norepinephrine reuptake inhibitor/SN-RI, serotonin norepinephrine reuptake inhibitor

One final comment about the SNRIs, they are SNDRIs; and the "D" stands for dopamine. In the human brain's dorsolateral prefrontal cortex, there is a paucity of dopamine relative to other parts of the brain. In this area, the brain will not waste nutrients to build a separate pathway/reuptake transporter for such a small amount of dopamine to return to the terminal button. Instead, dopamine hitches a ride inside the terminal button, riding piggyback on norepinephrine molecules. Dopamine and norepinephrine share transporters in the prefrontal cortex. When norepinephrine reuptake is blocked in the dorsolateral prefrontal cortex, dopamine cannot leave the synaptic cleft. A SNRI drug increases dopamine in the cleft by indirectly blocking its reuptake, too. All SNRIs are SNDRIs (Stahl, 2008).

Something Blue. Back to where it all started. Remember Alan Broadhurst, who in 1949, arrived at Geigy in Rhodes, England, to create a unique medication? Recall,

---

[16] Mechanism of Action (MOA)/not to be confused with Monoamine Oxidase (MAO)
[17] Neuroscience-based Nomenclature from *Stahl's Essential Psychopharmacology Prescriber's Guide*, 6th Edition.

he created 42 different derivatives of iminodibenzyl? Broadhurst tried the least toxic medication, G22150, which became Tofranil®/imipramine. Imipramine is an iminodibenzyl, a dye which goes by another name, "summer blue" or "sky blue" (Healy, 2002). In this case, the "something old" TCA is actually "something blue." The very first medication to treat the "blues" actually is a blue.

The SNRIs are a pharmaceutical story of accidentally discovering "something old," chemically turning them into "something borrowed," leading the way and morphing it 30 years later into "something new," and finally returning to its roots from whence all pharmaceuticals sprang, the humble dye. In this case, something blue.

QUIZ TIME

(1) Which of the following is NOT a SNRI?

  (a) Luvox®/fluvoxamine

  (b) Cymbalta®/duloxetine

  (c) Pristiq®/desvenlafaxine

  (d) Fetzima®/levomilnacipran

(2) Which of the following is NOT matched correctly, brand name/generic name?

  (a) Prozac®/escitalopram

  (b) Paxil®/paroxetine

  (c) Celexa®/citalopram

  (d) Viibryd®/vilazodone

(3) Which of the following SNRIs is thirty times more potent for serotonin reuptake blockade than norepinephrine reuptake blockade?

  (a) Effexor®/venlafaxine

  (b) Pristiq®/desvenlafaxine

  (c) Fetzima®/levomilnacipran

  (d) Cymbalta®/duloxetine

(4) Which of the drugs below represents the left-side enantiomer of the drug molecule?

  (a) Fetzima®/levomilnacipran

  (b) Lexapro®/escitalopram

  (c) Dexedrine®/d-amphetamine

  (d) a & b

(5) Why is it important to know both the brand and generic names?

  (a) Patients may know one name, and the prescriber may use the other.

  (b) Sometimes either the brand or the generic name is more prevalent in common parlance, but there is no rhyme or reason as to why this happens.

  (c) Sometimes there are multiple brand names for a generic drug (e.g., Advil® and Motrin® are brand names for generic ibuprofen). Knowing all names can prevent a patient from taking too much of the same medication.

  (d) All the above, and it keeps a prescriber on their toes.

BONUS QUESTION #1

True or False? Because SNRIs contain norepinephrine, these drugs have been known to throw a patient into a hypertensive crisis.

BONUS QUESTION #2

True or False? The serotonin ratio to norepinephrine in serotonin norepinephrine reuptake inhibitors (SNRIs) is approximately one to one, with reuptake blockade of serotonin and norepinephrine equally balanced.

ANSWERS

(1) The correct answer is "a," Luvox®/fluvoxamine. It is a SSRI that is used, almost exclusively, to treat obsessive compulsive disorder (OCD).

(2) The correct answer is "a." Prozac® is the brand, and its correct generic name is fluoxetine. The other choices are matched correctly.

(3) The correct answer is "a," Effexor®/venlafaxine. While Pristiq®/desvenlafaxine and Cymbalta®/duloxetine are also more potent for serotonin (as opposed to norepinephrine), it is the original Effexor®/venlafaxine, where the ratio is 30:1, serotonin:norepinephrine.

(4) The correct answer is "d," both choices "a" and "b." The giveaway is "l" and "levo," (along with "es") which are identifiers for "left side." The name of the generic medication in choice "c," d-amphetamine, begins with "d." This stands for "dextro," meaning right side of the molecule.

(5) The correct answer is "d," "all the above." Knowing all possible brand names along with the generic name adds another level of patient safety.

BONUS QUESTION ANSWER #1: TRUE. These drugs should be avoided if a patient has hypertensive and/or cardiac issues.

BONUS QUESTION ANSWER #2: FALSE. Effexor®/venlafaxine is approximately 30:1, serotonin to norepinephrine; Pristiq®/desvenlafaxine is approximately 11:1 in favor of serotonin; Cymbalta®/duloxetine is 9:1, serotonin to norepinephrine; and Fetzima®/levomilnacipran with a more balanced 1:2 ratio, in favor of norepinephrine. Do not assume that serotonin norepinephrine reuptake inhibitors equally block serotonin and norepinephrine.

Module 41

Neither Fish Nor Fowl

Atypical Antidepressants

☺ ☺ ☺

Just when you think it is possible to understand the usual terms that describe a drug's mechanism of action, drugs emerge that do not play by the standard "name-game rules." As a result, this topic is the poster child that poignantly expresses the need for standard nomenclature rather than the mishmash of historical descriptors which do not completely explain what is going on with a medication.

This class of antidepressants is called "atypical." There is no clear definition as to the identifying criteria for this collection of antidepressants. It is mostly inclusion by exclusion. They are neither pure SSRIs nor SNRIs, so they are something else, "atypical" and are not to be confused with "atypical antipsychotics."

These atypical antidepressants are a jumble of reuptake inhibitors, 5-HT$_{1A}$ partial agonists, and 5-HT$_{2A}$ antagonists. They are an assortment of medications that are not completely "pure" in their mechanism of action (e.g., a SRI, NRI, or SNDRI). For example, should Viibryd®/vilazodone be included with the SRIs or considered an atypical, since it is an "RI" plus something considered atypical, a partial agonist. Viibryd®/vilazodone is usually placed in the SSRI class. Then there is the newest one, Trintellix®/vortioxetine.

Since there is no clear agreement among professionals as to what an atypical antidepressant "is," the need for standard nomenclature among all psychotropics is painfully apparent. Given a quasi-textbook consensus, and admitting that others may be included in this class, the following medications are covered in this module and considered atypical: Desyrel®/trazodone, Wellbutrin®/bupropion, Remeron®/mirtazapine, and Trintellix®/vortioxetine.

Desyrel®/trazodone is a medication that most people know by its generic name, trazodone. It is FDA approved to treat depression and antagonizes the 5-HT_{2A} and 5-HT_{2C} receptors (a serotonin multimodal by NbN terminology). Few prescribers neither know its two brand names (Desyrel® and Oleptro®, extended release) nor use it as an antidepressant. At high doses it works as an antidepressant; however, at lower doses it is a sleep-inducing agent or sleep-cycle stabilizing agent. To that point, it is typically used off-label as a treatment for primary and secondary insomnia. This is a classic example of how the old mechanism of action (SARI, serotonin antagonist reuptake inhibitor) makes no mention of its sedative properties.

It produces sleep because it blocks the Histamine number one (H_1) receptor; in other words, it is an antihistamine. Since it is not dependent forming, it is preferred as a treatment for insomnia. It has one big sexual side effect: Risk of priapism (a prolonged, painful erection which, courtesy of erectile dysfunction (ED) commercials, every adult knows lasts longer than four hours).

Wellbutrin®/bupropion's sexist street nickname is "The Female Viagra." It provides "energy and pleasure" with no serotonin to harsh the desire for intimacy for both men and women. Its mechanism of action is a norepinephrine dopamine reuptake inhibitor. Even though it is activating, it does not improve sexual performance, just the expectation of it. In addition to NDRI, this medication has also been referred to as a "DRI," dopamine reuptake inhibitor and also a "CRI," catecholamine reuptake inhibitor.

Wellbutrin®/bupropion is FDA approved to treat major depressive disorder, seasonal affective disorder, and nicotine addiction. For depression, this medication impacts energy and mood, yet is known for its low switch or "flip" rate to mania. For smoking cessation, it is thought to function as a dopamine reuptake inhibitor (DRI) which occupies the dopamine transporter (DAT) in the striatum (caudate and putamen) and nucleus accumbens (NAc) and reduces cravings by mitigating the effects of nicotine withdrawal. Since there is a dopamine deficiency during nicotine withdrawal, Wellbutrin®/bupropion (more specifically Zyban®/bupropion SR which is FDA approved for smoking addiction) increases dopamine to offset the withdrawal symptoms. One important caveat concerning Wellbutrin®/bupropion: There is a warning for seizures at doses greater than 450 mg; however, seizures have also been reported at lower doses.

Remeron®/mirtazapine, covered in a previous module, was the treatment drug of choice for the patient dealing with both insomnia and depression. Remeron®/mirtazapine at lower doses is a sleep aid, and at higher doses it is an antidepressant. Its mechanism of action is NaSSA – noradrenergic specific serotonergic agent or norepinephrine antagonist serotonin specific antagonist. Remeron®/mirtazapine is FDA approved to treat major depressive disorder.

Approximately 16% discontinue use due to intolerable side effects – notably weight gain. Flu-like symptoms may emerge suggesting low white blood cell or granulocyte count (neutropenia) leading to increased susceptibility for infections. This drug blocks the alpha-2 (α_2) autoreceptor of the sympathetic nervous system. Because it is an autoreceptor and works backwards, blocking this receptor stimulates the release of norepinephrine. That explains the "N" part of the "NaSSA" mechanism of action. The "SSA" part refers to the fact that Remeron®/mirtazapine stimulates (agonizes) the $5\text{-}HT_{1A}$ receptor. Of the fourteen 5-HT (serotonin) receptors, $5\text{-}HT_{1A}$ is thought to have the most therapeutic effect on depression. When $5\text{-}HT_{2A}$ is blocked, many expected side effects are controlled, especially sexual ones.

Given all this agonizing and antagonizing, the old terminology for this drug should be "NaSSASa." That would read: Norepinephrine antagonist, serotonin specific agonist, and serotonin antagonist. Even the new NbN terminology (SN-Ran, serotonin, norepinephrine receptor antagonist) does not completely capture "what this drug does." Admittedly, knowing a drug's specific mechanism of action is cumbersome, but understanding it makes for a better prescriber, one who can match the problem with receptor while minimizing side effects. Put it all together, Remeron®/mirtazapine promotes more energy and the desired lifting of depression, with minimal to no sexual side effects; but look out for weight gain and flu-like symptoms.

Trintellix®/vortioxetine arrived on the US scene in 2013 and is "neither fish, nor fowl." Given the current terminology, it does not fit in any group and likely deserves its own category. The new NbN nomenclature better captures its mechanism of action and describes it as a S-MM, a serotonin drug that works multimodally on multiple 5-HT receptors with multiple actions. In that regard, it is certainly a unique antidepressant.

While there is some limited reuptake mechanism, its advantage lies in the fact that it explicitly targets only five of the fourteen serotonin receptors: 5-HT$_{1A}$ agonist, 5-HT$_{1B}$ partial agonist, 5-HT$_{1D}$ antagonist, 5-HT$_3$ antagonist, and 5-HT$_7$ antagonist. When a patient takes a SRI (serotonin reuptake inhibitor), it is an indirect agonist and agonizes indiscriminately all fourteen of the SRI receptors. This yields a myriad of side effects.

Trintellix®/vortioxetine, while not perfect, is far more selective, in that it either agonizes or antagonizes approximately one-third (five) of the fourteen serotonin receptors. Theoretically, it is thought that this combination of receptors, working synergistically, promotes the reduction of depressive symptoms and enhances cognitive actions (Stahl, 2017). Targeting specifically the receptor that treats the problem and avoiding agonizing or antagonizing any other receptors that do not serve the treatment purpose, this medication may reflect more of the future treatment of depression.

Trintellix®/vortioxetine is currently FDA approved for the treatment of major depressive disorder. It has a long half-life (t½ = 66 hours). It is a substrate, metabolized by CYP450 2D6, 3A4, 3A5, 2C19, 2C9, 2A6, 2C8, and 2B6. Regrettably, patients have reported more nausea with this medication. While no weight gain was reported in clinical trials, anecdotally, patients have acknowledged weight gain. Finally, this drug demonstrates potential advantages in that it expresses less sexual dysfunction, improves cognitive symptoms of depression, appropriate for elderly patients (however, given the CYP enzymatic action, approach cautiously), and is a possible treatment for patients who have not responded to antidepressants (Stahl, 2017).

The members of this group of "atypicals," do not fit easily into any category, yet they have demonstrated effectiveness since their introduction in the 1980s. This class exemplifies an evolution from general action to specific receptor action.

ATYPICAL ANTIDEPRESSANTS		
Brand (Year Approved in US)	Generic	MOA[18]/NbN[19]
Desyrel® (1981)	trazodone	SARI, 5-HT$_{2A}$ antagonist reuptake inhibitor/ S-MM, serotonin multimodal
Wellbutrin® (1985)	bupropion	NDRI, norepinephrine dopamine reuptake inhibitor/ D-RIRe, dopamine reuptake inhibitor and releaser
Remeron® (1996)	mirtazapine	NaSSA, noradrenergic and specific serotonergic agent/ SN-Ran, serotonin, norepinephrine receptor antagonist; the NbN-2 app states norepinephrine, serotonin multimodal[20]
Trintellix® (2013, with original name Brintellix®; 2016 name changed to Trintellix®)	vortioxetine	Multi-specific 5-HT receptor agonists and antagonists/ S-MM, serotonin multimodal

---

[18] Mechanism of Action (MOA)/not to be confused with Monoamine Oxidase (MAO)
[19] Neuroscience-based Nomenclature (NbN) from *Stahl's Essential Psychopharmacology Prescriber's Guide*, 6th Edition.
[20] Per NbN-2 app

QUIZ TIME

(1) The "atypical antidepressants" are considered "atypical" because _____.

    (a) they do not fit neatly into any one category

    (b) most are multimodal

    (c) they are a combination of agonists, antagonists, partial agonists, and indirect agonists

    (d) all the above

(2) True or False? Desyrel®/trazodone is known for its "off-label" use as a sleep aid rather than an antidepressant.

MATCH THE BRAND TO THE GENERIC

(3) _____Effexor®		A. imipramine
(4) _____Pristiq®		B. vortioxetine
(5) _____Cymbalta®		C. venlafaxine
(6) _____Fetzima®		D. desvenlafaxine
(7) _____Tofranil®		E. duloxetine
(8) _____Pamelor®		F. levomilnacipran
(9) _____Remeron®		G. nortriptyline
(10) _____Trintellix®		H. mirtazapine

ANSWERS

(1) The correct answer is "d," all the above. It is their uniqueness that places them outside the realm of standard antidepressants.

(2) TRUE. Desyrel®/trazodone is an antidepressant at higher doses, but it is used predominantly as a treatment for insomnia because it tends to stabilize the sleep cycle and facilitate sleep architecture.

(3) Effexor®/C, venlafaxine

(4) Pristiq®/D, desvenlafaxine

(5) Cymbalta®/E, duloxetine

(6) Fetzima®/F, levomilnacipran

(7) Tofranil®/A, imipramine

(8) Pamelor®/G, nortriptyline

(9) Remeron®/H, mirtazapine

(10) Trintellix®/B, vortioxetine

Module 42

Fourteen Drugs

Antipsychotic Medications

☺☺

Is it typical or atypical? Translated: Is it an *old*, typical antipsychotic, or is it a *new*, atypical antipsychotic? To answer that question, one must look back to the 1950s. There were no effective treatments for psychotic patients before that time – other than Adolf von Baeyer's 1904 barbiturates. They quieted the patient and made them easier to manage in the asylums. Because this medication was so calming, it obviated any need to further develop other medications that could have been more effective. In 1955 there were more than half-million psychotic patients in the United States, warehoused in mental institutions (Advokat, et al., 2014). From 1956 until 1983, the population dropped to 220,000 institutionalized patients. The creation of a dye, methylene blue, which sparked the creation of Thorazine®/chlorpromazine, along with President John F. Kennedy's 1963 Community Mental Health Act, shifted funds from institutions to community-based mental health treatment or care. This resulted in a dramatic decrease in institutionalized patient load, and patient care shifted to communities.

The first of the "typicals" was Thorazine®/chlorpromazine. According to research psychiatrists Jean Delay and Pierre Deniker, it created a "calmness, conscious sedation, and lack of interest in the detachment for external stimuli," and they referred to this as a "neuroleptic state" (Advokat, et al., 2014). This term, neuroleptic, some thought too scary, did not catch on in the United States where psychiatrists preferred "major tranquilizer," coined in the 1950s by F.F. Yonkman. A major tranquilizer was originally related to reserpine's calming effects (from the roots of *Rauwolfia serpentina* and *Rauwolfia vomitoria* plants). From neuroleptic to major tranquilizer, these drugs took time to metamorphosize into "typical antipsychotics."

In the mid-1960s, the second typical, Haldol®/haloperidol, emerged. Other typicals were developed and available today, but Thorazine®/chlorpromazine and Haldol®/haloperidol (even given a nickname, "Vitamin H" for Haldol®/haloperidol) are historically the most widely known and regularly used. If a medication is a typical antipsychotic, also referred to as a first-generation antipsychotic, it is more effective for treating positive symptoms of schizophrenia (e.g., hallucinations and delusions). Typicals are D_2 antagonists (block the dopamine number 2 receptor) and have an array of unwanted side effects, notably movement disorders (e.g., extrapyramidal symptoms and tardive dyskinesia).

One of the biggest differences between Thorazine®/chlorpromazine and Haldol®/haloperidol is potency. Thorazine®/chlorpromazine is less potent. Note the following chart comparing potency and expected side effects. This chart only applies to the typicals and is only a general reference, but it provides a criterion point concerning the overall expected side effect profile of the typicals.

TYPICAL ANTIPSYCHOTIC MEDICATIONS AND CORRESPONDING SIDE EFFECTS

DRUG	Potency	Anti-Cholinergic (DUCTS)[21]	Extra-Pyramidal Movements	Sedation	Orthostatic Hypotension (OH)
Thorazine®/ chlorpromazine	↓	↑	↓	↑	↑
Haldol®/ haloperidol	↑	↓	↑	↓	↓

Thorazine®/chlorpromazine is a low-potency typical, first-generation. With a low-potency typical, expect increased anticholinergic side effects (e.g., DUCTS) and a decreased probability of extrapyramidal movement disorders. But still very possible. Sedation and orthostatic hypotension (dizziness upon standing) will always be opposite the potency arrow. In the case of lower-potency Thorazine®/chlorpromazine, sedation and dizziness upon standing (Orthostatic Hypotension, OH) are anticipated side effects.

With Haldol®/haloperidol, a high-potency first-generation typical antipsychotic, the arrows flip. Anticholinergic side effects (e.g., DUCTS) are

---

[21] DUCTS: Dry mouth, urinary retention, constipation, tachycardia, sedation

decreased, and movement disorders are expected. Since sedation and dizziness (OH) follow the anticholinergic arrow, they are not as pronounced.

Some seventy years later, the typicals, first-generation antipsychotics (FGAs) are widely used, perhaps the most widely used for schizophrenia. They are also used "for other purposes, such as to treat nausea and vomiting, to sedate patients before anesthesia, to delay ejaculation, to relieve severe itching, to cure hiccups, and to manage psychosis regardless of cause (such as acute manic attacks, alcoholic hallucinosis, and psychedelic agents)" (Advokat, et al., 2014, p. 344).

TYPICAL AND ATYPICAL ANTIPSYCHOTICS
First- and Second-Generation[22]

First-Generation Typicals		
Brand (Year Entered US Market and/or Approved by FDA)	Generic	MOA[23]/NbN[24]
Thorazine® (1954)	chlorpromazine	D_2 antagonist/ DS-RAn, dopamine serotonin receptor antagonist
Haldol® (1958, first synthesized in Belgium; approved in 1967)	haloperidol	D_2 antagonist/ D-RAn, dopamine receptor antagonist

---

[22] Not a complete list
[23] Mechanism of Action (MOA)/not to be confused with Monoamine Oxidase (MAO)
[24] Neuroscience-based Nomenclature from *Stahl's Essential Psychopharmacology Prescriber's Guide*, 6th Edition.

Second-Generation Atypicals		
Brand (Year Entered US Market and/or Approved by FDA)	Generic	MOA[23]/NbN[24]
Clozaril®[25 & 26] (1959 initial studies conducted/1990)	clozapine	SDA, serotonin dopamine antagonist/ DSN-RAn, dopamine serotonin norepinephrine receptor antagonist
Geodon®[26] (2001)	ziprasidone	SDA, serotonin dopamine antagonist/ DS-RAn, dopamine serotonin receptor antagonist
Risperdal®[26] (1993)	risperidone	SDA, serotonin dopamine antagonist/ DSN-RAn, dopamine serotonin norepinephrine receptor antagonist
Zyprexa®[26] (1996)	olanzapine	SDA, serotonin dopamine antagonist/ DS-RAn, dopamine serotonin receptor antagonist
Seroquel®[26] (1997)	quetiapine	SDA, serotonin dopamine antagonist/ DS-MM dopamine, serotonin multimodal; the NbN-2 app includes norepinephrine[27]
Saphris®[26] (2009)	asenapine	SDA, serotonin dopamine antagonist/ DS-RAn, dopamine serotonin receptor antagonist
Latuda®[26] (2010)	lurasidone	SDA, serotonin dopamine antagonist/ DS-RAn, dopamine serotonin receptor antagonist; the NbN-2 app includes norepinephrine[27]
Invega®[26] (2009)	paliperidone	SDA, serotonin dopamine antagonist/ DS-RAn, dopamine serotonin receptor antagonist; the NbN-2 app includes norepinephrine[27]
Fanapt®[26] (2009, available 2010)	iloperidone	SDA, serotonin dopamine antagonist/ DS-RAn, dopamine serotonin receptor antagonist

---

[25] First developed as a "typical" in 1959; re-entered the US market as an atypical in 1990
[26] Also a mood stabilizer
[27] Partial mood stabilizer

Third-Generation		
Brand (Year Entered US Market and/or Approved by FDA)	Generic	MOA[23]/NbN[24]
Abilify®[26] (2002)	aripiprazole	dopamine partial agonist/ DS-RPA, dopamine serotonin receptor partial agonist
Rexulti®[28] (2015)	brexpiprazole	dopamine partial agonist/ DS-RPA, dopamine serotonin receptor partial agonist
Vraylar®[27 & 29] (2015)	cariprazine	dopamine partial agonist/ DSN-Ran, Dopamine, Serotonin, norepinephrine receptor antagonist; the NbN-2 app omits Norepinephrine[8]

For years, no new antipsychotics were marketed in the United States. While the first-generation medications emerged in the 1950s golden age, atypicals emerged in the 1980s and represented a dramatic research breakthrough. The result was a revolution that offered improvements relative to side effects. The atypicals, often referred to as second-generation antipsychotics (SGAs), are undoubtedly different from the old typicals; but, unfortunately, they may not be any more effective.

Second-generation antipsychotics appeared contiguous with the ability to develop medications based on receptor profiling - specifically radio-labeled binding technologies by American neuroscientist Solomon Halbert Snyder. Because of his linking specific neurotransmitters and drugs and clarifying psychotropics' action to their receptors, Snyder is credited with numerous advances in molecular neuroscience (Healy, 2002). By the late 1980s, producing a drug that was both a dopamine and

---

[28] Per the NbN-2 App
[29] Sometimes included as a second-generation antipsychotic

serotonin antagonist was routine, and this became the molecular basis for atypical, second-generation antipsychotics. Without Snyder's groundbreaking research findings and concomitant advances, second-generation medications would likely not exist. Although the chapter on opioids is forthcoming, Snyder is most famous for his research on the opioid receptors. He is one of the most cited researchers globally; his work is vastly admired; he is deserving of much greater name recognition.

Snyder's "radio-labeled binding" distinguished between 5-HT_1 and 5-HT_2 receptors, creating a logical area for drug research and development. Pharmaceutical companies, Janssen and Eli Lilly, embraced these new receptors and demonstrated that drugs which agonize serotonin could produce psychotogenic effects and worsen psychoses. Therefore, serotonin antagonism was a logical research inquiry with the hope to improve psychotic symptomology. The rest is history; it was proved true, and nowhere is this better demonstrated than an old, yet new again medication — Clozaril®/clozapine.

The newer "atypicals" are serotonin dopamine antagonists or "SDAs." There is an interesting twist of history concerning a drug that was initially a first-generation, typical neuroleptic that became today's Clozaril®/clozapine. It was first developed as a neuroleptic in 1959, but it did not catch on. In the 1950s, it was believed that major tranquilizers, typicals, first-generation medications must create side effects, notably movement disorders (known as the "Theory of Neuroleptic Action"), before they were deemed effective. Although Clozaril®/clozapine produced akathisia and neuroleptic malignant syndrome (NMS), it did not create the classic extrapyramidal side effects (Healy, 2002). Given the medical beliefs then, it was deemed ineffectual and useless.

Clozaril®/clozapine was used in Europe where its lack of movement disorders and extrapyramidal side effects was appreciated. However, it had a darker side — the probability of agranulocytosis (loss of white blood cells in the blood and decreased immunity) emerged. With this worry, the drug was withdrawn from unrestricted use in Europe. This medical issue continues to haunt Clozaril®/clozapine to this day.

Clozaril®/clozapine was shelved for years. Finally, it was "dusted off" and re-entered the modern 1990s pharmacopeia. It was re-examined because "(1) agranulocytosis was found to be reversible when the drug was discontinued, and (2) the drug was found to be therapeutically beneficial in patients who had failed to respond to the traditional antipsychotic medications" (Advokat, et al., 2014, p. 352). As a result, "Clozapine was a major advance because it was effective for many patients (about 30%) who did not respond to standard treatment and because it produced little or no movement disorders such as EPS or TD (and in fact, may even reduce TD caused by other antipsychotics)" (Advokat, et al., 2014, p. 341). It may also reduce the risk of suicide in patients with schizophrenia relative to other antipsychotics.

Between 1997 and 2011, atypical antipsychotics experienced a three-fold increase in the number of overall antipsychotics prescribed in the United States (not only in the US, but similar trends were reported in Canada and Europe), exposing at least five million people to antipsychotics each year. Of that number, 20% of nursing home residents are being prescribed antipsychotics annually (Factor, et al., 2019). Additionally, this overprescribing trend extends to children and adolescents. A ten-fold increase was reported among patients between the ages of four and eighteen between 1993 and 2010, and 65% of those were prescribed off-label (Factor, et al., 2019). The second-generation, atypical antipsychotics are FDA approved to treat a variety of disorders including, but not limited to, schizophrenia, types of manic-depressive disorders, psychoses, and others.

ANTIPSYCHOTICS
1ST, 2ND, 3RD Generation

"Neuroleptics," First-Generation Typicals
(not a complete list)

Brand (Year Entered US Market)	Generic	FDA Approved Treatments[30]
Thorazine® (1950)	chlorpromazine	Schizophrenia, psychosis, nausea, vomiting, restlessness and apprehension before surgery, acute intermittent porphyria, manifestations of manic type of manic-depressive illness, tetanus (adjunct), intractable hiccups, combativeness and/or explosive hyperexcitable behavior (in children), hyperactive children who show excessive motor activity with accompanying conduct disorders consisting of some or all of the following symptoms: impulsivity, difficulty sustaining attention, aggressivity, mood lability, and poor frustration tolerance
Haldol® (1958)	haloperidol	Manifestations of psychotic disorders (oral, immediate-release injection), tics and vocal utterances in Tourette's syndrome (oral, immediate-release injection), second-line treatment of severe behavior problems in children of combative, explosive hyperexcitability (oral, immediate-release injection), second-line short-term treatment of hyperactive children (oral, immediate-release injection), treatment of schizophrenic patients who require prolonged parenteral antipsychotic therapy (depot intramuscular decanoate)

---

[30] Per *Stahl's Essential Psychopharmacology Prescriber's Guide*, 6th Edition.

Second-Generation Atypicals

Brand (Year Entered US Market)	Generic	FDA Approved Treatments[30]
Clozaril® (1959/1990)	clozapine	Treatment-resistant schizophrenia, reduction in risk of recurrent suicidal behavior in patients with schizophrenia or schizoaffective disorder
Geodon® (2001)	ziprasidone	Schizophrenia, delaying relapse in schizophrenia, acute agitation in schizophrenia (intramuscular), acute mania/mixed mania, bipolar maintenance
Risperdal® (1993)	risperidone	Schizophrenia, ages 13 and older (oral, long-acting microspheres intramuscularly); delaying relapse in schizophrenia (oral); other psychotic disorders (oral); acute mania/mixed mania, ages 10 and older (oral, monotherapy and adjunct to lithium or valproate); autism-related irritability in children ages 5-16; bipolar maintenance (long-acting microspheres intramuscularly monotherapy and adjunct to lithium or valproate)
Zyprexa® (1996)	olanzapine	Schizophrenia (ages 13 and older); maintaining response in schizophrenia; acute agitation associated with schizophrenia (intramuscular); acute mania/mixed mania (monotherapy and adjunct to lithium or valproate) (ages 13 and older); Bipolar maintenance; acute agitation associated with bipolar I mania (intramuscular); bipolar depression in combination with fluoxetine (Symbyax®); treatment-resistant depression in combination fluoxetine (Symbyax®)
Seroquel® (1997)	quetiapine	Acute schizophrenia in adults (quetiapine, quetiapine XR) and ages 13-17 (quetiapine); schizophrenia maintenance (quetiapine XR); acute mania in adults (quetiapine and quetiapine XR, monotherapy and adjunct to lithium or valproate) and ages 10-17 (quetiapine, monotherapy and adjunct to lithium or valproate); bipolar maintenance (quetiapine, quetiapine XR); bipolar depression (quetiapine, quetiapine XR); depression (quetiapine XR, adjunct)
Saphris® (2009)	asenapine	Schizophrenia, acute and maintenance (adults); acute mania/mixed mania, monotherapy (ages 10 to 17 and in adults); acute mania/mixed mania, adjunct to lithium or valproate (adults)
Latuda® (2010)	lurasidone	Schizophrenia (ages 13 and older) Bipolar depression
Invega® (2006)	paliperidone	Schizophrenia (ages 12 and older); maintaining response in schizophrenia; schizoaffective disorder
Fanapt® (2009, available 2010)	iloperidone	Schizophrenia Schizophrenia maintenance

Third-Generation		
Brand (Year Entered US Market)	Generic	FDA Approved Treatments[30]
Abilify® (2002)	aripiprazole	Schizophrenia (ages 13 and older) (Abilify®, Abilify Maintena®, Aristada®); maintaining stability in schizophrenia; acute mania/mixed mania (ages ten and older; monotherapy and adjunct); bipolar maintenance (monotherapy and adjunct); depression (adjunct); autism-related irritability in children ages 6 to 17; Tourette's disorder in children ages 6 to 18; acute agitation associated with schizophrenia or bipolar disorder (IM)
Rexulti® (2015)	brexpiprazole	Schizophrenia Treatment-resistant depression (adjunct)
Vraylar® (2015)	cariprazine	Schizophrenia Acute mania/mixed mania

It is thought that the main difference between first- and second-generation antipsychotics is the likelihood that typicals (1st generation) cause movement disorders, and atypicals (2nd generation) do not. That assumption is likely insufficient to account for the naming difference. Another possible reason for the difference is that typicals only block dopamine with affinity, and atypicals block both dopamine and serotonin. However, there is possibly another, more valid reason: the dissociation constant.

The dissociation constant (K_d) is how rapidly a drug attaches to and disengages from a receptor. The old, first-generation typicals tend to bind and "hold on" to their target. Admittedly this binding is milliseconds, but neurotransmitters attach and dissociate (release) slower than the atypicals. Atypicals, the second-generation bind and release quickly, certainly more quickly than the old first-generation antipsychotics. Additionally, there are drugs and/or other chemicals that irreversibly bind and never let go until the receptor dies. Examples include old monoamine oxidase inhibitors (MAOIs) and organophosphates (e.g., nerve gases including sarin, soman, Novichok, and VX).

The third-generation antipsychotics received their moniker for lack of a more accurate, better name. They are different from first-generation typical antipsychotics; they are consistently different from the serotonin dopamine antagonist (SDA), second-generation, atypical antipsychotics; and tweaked enough to garner their own unique label – third-generation. Instead of complete agonists and antagonists, they tend to be partial agonists. That said, Abilify®/aripiprazole and Rexulti®/brexpiprazole, fit that definition; but Vraylar®/cariprazine is more accurately a second-generation. Because of its late entry into the market, it is considered a third-generation, even a mood stabilizer. To make it more confusing, according to the old nomenclature, Vraylar®/cariprazine is a dopamine partial agonist; however, with the new NbN terminology, it is a DSN-Ran, dopamine serotonin norepinephrine receptor antagonist (Stahl, 2017).

What is astounding and extraordinary is the range and scope of mental illnesses that these fourteen drugs (typicals, atypicals, and third-generation) treat. Albeit helpful and foundational medications, their exact mechanism of action continues to be unknown; notwithstanding, future medications will likely build on the past and hold both promise and cure.

QUIZ TIME

(1) Which of the following drug pairs are considered typicals, first-generation antipsychotics?

(a) Rexulti®/brexpiprazole & Abilify®/aripiprazole

(b) Luvox®/fluvoxamine & Zoloft®/sertraline

(c) Thorazine®/chlorpromazine & Haldol®/haloperidol

(d) Thorazine®/chlorpromazine & Risperdal®/risperidone

(2) Which of the following typical drugs was originally a dye, methylene blue, before it became an antipsychotic?

(a) Tofranil®/imipramine

(b) Vraylar®/cariprazine

(c) Risperdal®/risperidone

(d) Thorazine®/chlorpromazine

(3) The first, second, and third-generation antipsychotics share one treatment modality in common: Their ability to treat _____.

(a) schizophrenia

(b) personality disorders

(c) bipolar disorders

(d) seizure disorders

(4) If a first-generation, typical antipsychotic has anticholinergic side effects in its profile, chances are it is _____.

(a) high potency

(b) low potency

(c) high efficacy

(d) low efficacy

(5) In the future, it is hoped that antipsychotics will have _____.

(a) fewer side effects

(b) clearer mechanism of action

(c) less drug-drug interactions

(d) all the above

ANSWERS

(1) The correct answer is "c," Thorazine®/chlorpromazine and Haldol®/haloperidol are the classic typical, first-generation antipsychotics.

(2) The correct answer is "d," Thorazine®/chlorpromazine was originally a dye, methylene blue, created by Heinrich Caro at BASF.

(3) The correct answer is "a," schizophrenia. All first, second, and third-generation antipsychotics are FDA approved to treat schizophrenia. Other drugs, approved by the FDA, are more receptor and disorder specific.

(4) The correct answer is "b," low potency. It is likely Thorazine®/chlorpromazine, a low potency antipsychotic known for its anticholinergic side effects (DUCTS: dry mouth, urinary retention, constipation, tachycardia, and sedation).

(5) The correct answer is "d," all the above. The future holds promise, clarity, and a possible cure.

Sweet Move on the QTc

Three Main Side Effects of Antipsychotics

☺

When it comes to antipsychotics – first, second, or third-generation – numerous drug-specific side effects require caution; but overall, there are three significant, often unrecognized problems, all too often inappropriately managed, which must be discussed. This section is presented with an overly broad brush. Not every antipsychotic causes these three issues; however, it is almost malpractice not to evaluate for them because they are extremely easy to identify during any clinical interview.

The antipsychotics can be unpredictable and full of exceptions. This antipsychotic can cause weight gain, but not heart issues. Another has a low probability of movement disorders but can cause agranulocytosis. Then there is the antipsychotic that at a low dose performs like the newer atypicals; but at higher doses, it turns into an old typical from the 1950s. Clinicians spend a lifetime learning the nuances of antipsychotic medications, which is certainly a worthy goal; however, it is a daunting task from the onset.

For the beginning student, remember this broad rule of thumb: Check the patient for all three of the following problems. If you catch something that is not supposed to happen – and unexpected things do happen – the patient is protected, which is the definition of good healthcare. These three issues are (1) movement disorders, (2) electrocardiographic abnormalities (changes in the electrical conductivity of the heart or increases in the QTc interval), and (3) diabetes and hyperglycemia.

Movement Disorders. One of the few truths in psychopharmacology is that old typical antipsychotics cause movement disorders. That is fact. However, drug representatives for the newer medications allege that the "modern" atypicals do not cause movement disorders. Nothing could be further from the truth. Granted, one is not as likely to see movement disorders with the newer atypicals, but movement disorders can happen when least expected. A recent meta-analysis on the prevalence of tardive dyskinesia (the more puzzling movement disorder) from 2000 to 2015 found prevalence was 25% and 30% for second-generation and first-generation antipsychotics, respectively (Carbon, et al., 2017).

If a drug manipulates dopamine, either agonizes or antagonizes, then screening for movement disorders should be part of any drug interview/drug check. The old typicals specifically block the D_2 receptor; this profoundly influences the nigrostriatal pathway (dopaminergic pathway responsible for movement). The newer atypicals block both serotonin and dopamine; as a result, blocking serotonin mitigates dopamine's impact on the nigrostriatal pathway, thus reducing the probability of movement disorders. The keyword is "probability." The stimulants agonize dopamine – watch for movement disorders. The SRIs increase serotonin and indirectly block dopamine – watch for movement disorders, explicitly biting the cheek. The rule is worthy of reiteration: Agonize or antagonize dopamine receptors, by any means, watch for movement disorders.

The broad syndrome of drug-induced movement disorders is referred to as extrapyramidal side effects. This name is derived from the fact that this disorder stems from a location outside the pyramidal tracts. The term "extra" distinguishes it from the motor cortex's pyramidal tracts that reach their targets by traveling through the pyramids of the medulla oblongata. Pyramidal tracts are concerned with voluntary movement; extrapyramidal neurons, outside the pyramidal system, are concerned with fine motor movements or the "fine-tuning" of involuntary movements.

Within this domain of extrapyramidal reactions, there are two types: (1) acute extrapyramidal reactions, which develop early in treatment in up to 90% of patients and (2) tardive (as in "tardy") dyskinesia, which occurs later, during, and even after cessation of antipsychotic treatment (Advokat, et al., 2014). There are three types of acute extrapyramidal specific side effects: (1) dystonia, (2) akathisia, and (3) neuroleptic-induced parkinsonism. The non-acute extrapyramidal reaction that is more puzzling, serious, and late occurring is tardive dyskinesia.

A dystonia is a medical emergency and can manifest within the first few hours of taking the medication. It is characterized by slow, sustained, involuntary muscular contractions and can involve the entire body or individual parts. One example is an oculogyric crisis, a prolonged involuntary upward deviation of the eyes, where the patient's eyes involuntarily roll back in their orbits.

Another dystonia example that is equally dramatic and painful to watch is torticollis. Derived from the Latin words, "tortus" for "twisted" and "collum" for "neck," the derivation of torticollis describes this condition perfectly. It is characterized by an abnormal, twisted, asymmetrical head or neck position.

The least recognized dystonia is laryngeal dystonia and is often confused with shortness of breath. It can be life-threatening and has been reported predominantly in young males (Christodoulou & Kalaitzi, 2008).

The usual definition taught in graduate school is that akathisia is a movement disorder typified by an inability to sit still. However, this emphasis on physical movements often neglects to point out that akathisia may also include "mental movements" (e.g., restless thoughts). Additionally, akathisia can also be a feeling of inner restlessness, distress, anxiety, and racing thoughts. From Greek, "kathis" means "sitting," the "a" prefix denotes "not," roughly translated, akathisia means "not sitting."

The term akathisia was coined at the turn of the century by Ladislav Haskovec; however, the condition had been reported as early as 1861 by physician Armand Trousseau who, upon watching one of Louis Napoleon's courtier's inability to sit still, determined that the problem was not a "willed one."

Akathisia reappeared with encephalitis lethargica (as identified by Italian physician Constantin von Economo, and made famous by Oliver Saks' 1973 book, *Awakenings*) in Europe's 1918 Great Influenza pandemic. For thirty years, the condition remained unrecorded until the phenomenon re-emerged with the creation of neuroleptics and their related movement disorders. While the exact cause was unknown and given the newly emerging field of "neurotransmitters," the cause of akathisia remained elusive, even after it was acknowledged that Thorazine®/chlorpromazine could induce it (Healy, 2002; Sachdev, 1995).

Finally, in the 1970s, the work of Philip May and Theodore van Putten once again drew attention to the insidious aspects of akathisia (van Putten, 1975). They noted that akathisia was often confused with tardive dyskinesia. Akathisia, given its distressing nature, may incite suicide and violence.

This is one of the reasons underlying the 2004 "Black Box Warning" (a warning given by the FDA to a drug when a particular side effect becomes so prevalent the FDA believes the population should be warned) issued to SRIs. This warning cautions the clinician to be aware of suicidal ideation, suicidal thinking, feeling, and behavior in people aged twenty-four and younger. SRIs can cause SRI-induced akathisia. When SRIs indirectly agonize serotonin, they also indirectly block the production of dopamine. Akathisia can present within four days after taking an antipsychotic. Fidgeting, rocking back and forth, and pacing are all physical symptoms. While the legs are the most often affected, "racing thoughts," inner restlessness, distress, and anxiety must all be included in the broad category of akathisia.

Neuroleptic-induced or drug-induced Parkinsonism resembles idiopathic (of unknown etiology) Parkinson's disease. It is characterized by Parkinson-like symptomatology including tremor at rest, "cogwheel" rigidity of limbs, slowing of movement, slurred speech, postural instability, difficulty maintaining posture with an unsteady gait, and reduction in spontaneous movements. Drug-induced Parkinsonism (DIP) which frequently produces disability in the elderly, has replaced tardive dyskinesia (TD) as the most significant neurological complication of antipsychotic drugs. Dramatically increasing the risk of falls and hospitalizations, this condition is an akinetic-rigid syndrome that mimics Parkinson's disease. Along with dystonia and akathisia, it is usually reversible.

Since the 1950s, when tested for its antipsychotic potential, reserpine was noted to cause a Parkinsonian-like state in humans. This serendipitous finding led to the discovery that dopamine, when significantly blocked and depleted to about 20% of normal levels, precipitates neuroleptic-induced Parkinson's disease. Symptoms emerge when antipsychotic blockade of dopamine receptors reaches 80% or greater (Carlson, 1959; Advokat, et al., 2014; Factor, et al., 2019).

Tardive dyskinesia is a severe movement disorder. Characterized by involuntary hyperkinetic movements (tongue, face, trunk, and extremities) which include smacking and pursing of lips, lateral jaw movements, darting/pushing/twisting of the tongue, choreiform (dancelike) movements, rocking of the trunk, pelvic thrusting, rotating ankles and legs, marching in place, and irregular respirations. Any of these can appear within a few months to several years after initiation of neuroleptic-induced Parkinsonism. While it is likely dose related, there appears to be substantial variation among individual patients. Most common in patients 50 or older, especially women, and approximately 50% of the elderly are affected after five years of drug use. Tardive dyskinesia is not as likely with atypical antipsychotics, but the incidence rate is certainly not zero.

Presently, there is no effective, adequate treatment for tardive dyskinesia. Clozaril®/clozapine may reduce symptomology, and recently a novel drug, Ingrezza®/valbenazine, was developed by Neurocrine Biosciences, Inc. Its mechanism of action is a vesicular monoamine transporter inhibitor (VMAT Inhibitor) and blocks the entrance of neurotransmitters into vesicles located inside the terminal button. At depolarization, when the vesicles release their contents, there is less to release, and this indirect treatment reduces dopamine in the cleft. This provides a steady level, albeit lower dopamine level, and minimizes dopamine depletion side effects (Advokat, et al., 2014).

To offer an additional explanation, chronic dopamine blockage over the years may cause postsynaptic D_2 receptors to up-regulate and become hypersensitive to dopamine. This means, when there is a prolonged lack of receptor stimulation due to reduced amounts of a neurotransmitter, the number of receptors will increase and become hypersensitive. This up-regulation takes place over weeks, not hours, and not all receptors show this effect. A hypersensitivity of D_2 receptors in the nigrostriatal pathway is thought to be the cause of tardive dyskinesia. By decreasing dopamine uptake into vesicles, the amount of dopamine available for release into the synapse interacting with hypersensitive D_2 receptors, is reduced and is thought to lead to a decrease in symptomology (Advokat, et al., 2019).

Electrocardiographic Abnormalities. Electrocardiographic abnormalities are a worthy concern relating to all antipsychotic medications. The heart has its own natural pacemaker, the sinoatrial node, located in the right atrium. An electrical signal is generated from this pacemaker and flows through the atria and into the ventricles through the atrioventricular node. The result: The ventricles of the heart contract; blood is propelled into the aorta and the arteries. After each heartbeat (depolarization and mechanical contraction), the heart resets, repolarizes, and the process begins anew. The QTc interval (one heartbeat and often referred to as QTc, with "c" standing for "corrected") is the time from the start, sinoatrial node impulse, ventricular depolarization, to ventricular repolarization. This QTc is one cycle, one heartbeat, an average of 80 times per minute, 115,000 times each day, 42 million times per year, and 3 billion times in a lifetime. The heart, throughout a lifetime, pumps about one million barrels of blood.

If the heartbeat time/interval, from start to finish, is prolonged by 500 milliseconds (1/2 of one second), there is a significant risk of heart arrhythmia (torsades de pointes, French for "twisting of points") that can result in death. Many psychotropics (antipsychotics, antidepressants, stimulants, and antianxiety/anxiolytics) can cause QTc intervals to lengthen. Among first-generation antipsychotics, Mellaril®/thioridazine (rarely prescribed) is most associated with this problem. Among the second-generation antipsychotics, Geodon®/ziprasidone is known for increasing the QTc (Advokat, 2014).

Diabetes and Hyperglycemia. The third problem area, diabetes and hyperglycemia, is also associated with the use of certain (not all) antipsychotics. Patients who receive atypical antipsychotics run the risk of developing adult-onset (type 2) diabetes (Schatzberg & DeBattista, 2019). The risk is high with Clozaril®/clozapine and Zyprexa®/olanzapine; medium risk with Risperdal®/risperidone and Seroquel®/quetiapine; and low with

Geodon®/ziprasidone and Abilify®/aripiprazole. When the patient states, "I'm always hungry." check the medications. This is a major clue that advanced evaluation is required.

Because treatment for schizophrenia places patients at possible increased risk for diabetes, all mental health professionals must be aware of "metabolic syndrome." Metabolic syndrome (also known as Syndrome X) is a frequent finding in type 1 diabetes. It is a clustering of metabolic risks including increased blood pressure and triglycerides, insulin resistance (cells become numb to insulin produced by the body and precedes the onset of type 2 diabetes), obesity (apple shape), decrease in high-density lipoprotein (HDL) cholesterol, and increased risk of cardiovascular disease (Lieberman, 2004).

Three important side effect considerations have been presented with an overly broad brush. Not all antipsychotics cause these problems, but a substantial majority can drive one, two, or all of them. Because typical, atypical, and third-generation antipsychotics combined with an individual's genetics can always surprise, it is wise to monitor for all three side effects no matter what antidepressant, antipsychotic, stimulant, other medication, and/or combination the patient is taking. Catch it when it is unexpected, and it is a win-win for the patient and the clinician.

QUIZ TIME

(1) Movement disorders are the result of blocking _____ receptors.

     (a) histamine

     (b) serotonin

     (c) adenosine

     (d) dopamine

(2) Movement disorders are the result of blocking the _____ pathway.

     (a) nigrostriatal

     (b) tuberoinfundibular/tuberohypophyseal

     (c) mesolimbic

     (d) incertohypothalamic

(3) Which of the following SNRIs is thirty times more potent for serotonin reuptake blockade, as opposed to norepinephrine reuptake blockade?

     (a) Effexor®/venlafaxine

     (b) Pristiq®/desvenlafaxine

     (c) Fetzima®/levomilnacipran

     (d) Cymbalta®/duloxetine

(4) A _____ is a medical emergency.

     (a) dystonia

     (b) akathisia

     (c) tardive dyskinesia

     (d) neuroleptic-induced parkinsonism

(5) Which of the following psychotropics can cause cardiovascular issues and/or lengthen the QTc interval?

     (a) antipsychotics

     (b) antidepressants

     (c) stimulants

     (d) antianxiety/anxiolytics

     (e) all the above

ANSWERS

(1) The correct answer is "d," dopamine. When dopamine is agonized or antagonized, expect movement disorders.

(2) The correct answer is "a," nigrostriatal.

(3) The correct answer is "a," Effexor®/venlafaxine. While Pristiq®/desvenlafaxine and Cymbalta®/duloxetine are also more potent for serotonin (as opposed to norepinephrine), it is the original Effexor®/venlafaxine, where the ratio is 30:1, serotonin:norepinephrine.

(4) The correct answer is "a," dystonia. For both the humanity of it and medical necessity, relief from these muscular spasms requires immediate attention.

(5) The correct answer is "e," all the above. Remember the rule of thumb: Monitor all psychotropics for movement disorders, heart issues, and diabetic concerns.

In the Mood

Mood Stabilizers

☺ ☺ ☺

Time for myth-busting. There is no such thing as a "mood stabilizer," although that is the name of this drug class. A more accurate term is "neuromodulator," but that has insufficient marketing zing. That said, the selection of a "mood stabilizer" must serve two goals: (1) Reduction of current symptoms and (2) relapse prevention. Mood disorders are known for their recurrence, so a prophylaxis treatment is likely necessary (Preston & Johnson, 2016). More often, both prescriber and patient are considered successful (and perhaps a little lucky) when treatment significantly reduces both the number and frequency of bipolar episodes.

From melancholia to manic depression to bipolar, mood stabilizers can treat both depression and manic symptoms. Some medications are quite good at treating depression, but not mania. Others are good at mania but not depression. Antipsychotics are considered the "big guns." However, it all started with lithium, the 3rd element on the Periodic Table.

In the 1st Century, one of the most renowned Greek physicians, Aretaeus of Cappadocia acknowledged depression and mania symptoms. He prescribed bath salts to calm manic symptoms, and no doubt lithium was the main ingredient in Greek bath salts. While the remedy is ancient, it was not until Australian physician John Cade rediscovered the remedy during the 1950s golden age. Working with guinea pigs, he noticed that lithium had a tranquilizing effect. He tried it on his patients; the results were mixed; some lived; some died. He discovered, the hard way, that lithium was toxic. Undaunted, he published his results; other prescribers gave it to their patients; most were unimpressed (Cade, 1949).

Danish researcher Mogens Schou did the first randomized trial and demonstrated lithium's (1) efficacy and (2) safe blood levels. To this day, blood draws are necessary to find lithium's "sweet spot," the non-toxic blood level.

Lithium's exact mechanism of action is unknown, complex, and a mystery; as a result, there is no old nomenclature to describe it. "Neuroscience-based Nomenclature (NbN)" defines it as "lithium enzyme interactions" (Li-Eint; the NbN-2 app calls it a "lithium enzyme modulator"). It likely inhibits second-messenger, inositol monophosphatase, affecting its transmission. Translated: Via second messenger, lithium slows things down, boosts transmission of GABA, and reduces excitatory neurotransmitters like dopamine and glutamate. It is FDA approved to treat manic episodes of manic-depression illness and maintenance treatment for manic-depressive patients with a history of mania. One unique aspect of lithium is its ability to decrease the risk of suicide (Schatzberg & DeBattista, 2019).

Eskalith®/lithium is an old drug, the third element of the Periodic Table and little money to be made from it. Nevertheless, lithium has recently resurged in popularity. Because lithium has a narrow therapeutic index, there is a small difference between the effective dose and the toxic dose. A blood test to check lithium levels should be performed close to 12 hours following the last dose to assure an accurate reading. The therapeutic range is 0.6-1.2 mmol/L (millimoles per liter); toxic level is 1.5 mmol/L; and severe toxicity level is 2.0 mmol/L.

Eskalith®/lithium has numerous side effects. The following mnemonic, "BATTERY" (actually "BAATTTERRY") suggests a partial list:

- Bradycardia
- Ataxia & Acne
- Tremor & hypoThyroid
- Teratogenic (first-trimester pregnancy risk)
- Edema
- Rash & Renal toxicity
- leukocYtosis (WBC above normal and reversible upon discontinuation)

Other mood-stabilizing, anti-manic agents can be used alone or combined with lithium. These FDA approved antipsychotics (partial listing) include Seroquel®/quetiapine, Latuda®/lurasidone®, Zyprexa®/olanzapine and Vraylar®/cariprazine; mood-stabilizing anticonvulsants (Depakote®/valproic acid, Tegretol®/carbamazepine, and Lamictal®/lamotrigine). Neurontin®/gabapentin, Topamax®/topiramate, and Gabitril®/tiagabine have numerous benefits, but randomized trials have demonstrated they are no better than placebo in the treatment of bipolar disorders (Berlin, et al., 2015; Kushner, et al, 2006).

BIPOLAR DISORDER MEDICATIONS: MANIA

Brand	Generic	MOA[31]/NbN[32]
Eskalith®, Lithonate®, Lithobid®	lithium[33]	Undefined/ Li-Eint, lithium enzyme interactions Li enzyme modulator[34]
Tegetrol®,Equetro®	carbamazepine	Undefined/ Glu-CB, glutamate, voltage-gated sodium and calcium channel blocker
Depakene®, Depacon®, Depakote®, Stavzor®	valproic acid	Undefined/ Glu, yet to be determined
Zyprexa®	olanzapine	SDA, serotonin dopamine antagonist/ DS-Ran, dopamine and serotonin receptor antagonist

BIPOLAR DISORDER MEDICATIONS: DEPRESSION

Brand	Generic	MOA[31]/NbN[32]
Seroquel®	quetiapine	SDA, serotonin dopamine antagonist/ DS-MM, dopamine, serotonin multimodal NbN-2 App dopamine, serotonin, norepinephrine, Multimodal[34]
Zyprexa®	olanzapine	SDA, serotonin dopamine antagonist/ DS-Ran, dopamine and serotonin receptor antagonist
Latuda®	lurasidone	SDA, serotonin dopamine antagonist/ DS-Ran, dopamine and serotonin receptor antagonist
Lamictal®	lamotrigine[35]	Voltage-sensitive sodium channel antagonist/Glu-CB, glutamate and voltage-gated sodium channel blocker

While these medications do not stabilize mood, they offer a means to modulate mood and must be monitored carefully and regularly. Of all the psychotropics, this is probably the best example of a condition, mood disorder, that must be diagnosed correctly, otherwise no modulation of any kind will work. It is also a lifetime commitment to avoid relapse.

---

[31] Mechanism of Action (MOA)/not to be confused with Monoamine Oxidase (MAO)
[32] Neuroscience-based Nomenclature from *Stahl's Essential Psychopharmacology Prescriber's Guide*, 6th Edition.
[33] FDA approved for mania and maintenance
[34] 5Per NbN-2 App
[35] 4Consider for rapid cyclers (Preston & Johnson, 2016)

QUIZ TIME

(1) Which of the following is NOT a SNRI?

    (a) Luvox®/fluvoxamine

    (b) Cymbalta®/duloxetine

    (c) Pristiq®/desvenlafaxine

    (d) Fetzima®/levomilnacipran

(2) Lithium, the 3rd element on the periodic table, is approved for _____.

    (a) mania

    (b) depression

    (c) maintenance

    (d) both "a" and "c"

(3) Which of the following atypicals is NOT approved for the treatment of bipolar depression?

    (a) Geodon®/ziprasidone

    (b) Seroquel®/quetiapine

    (c) Latuda®/lurasidone

    (d) Zyprexa®/olanzapine

(4) Which of the following is the therapeutic serum level of lithium?

    (a) 0.1 – 1.0 mmol/L

    (b) 0.6 – 1.2 mmol/L

    (c) 1.5 – 2.0 mmol/L

    (d) 2.1 – 2.7 mmol/L

(5) A blood test to determine lithium levels should be performed as close to _____ hours after the last dose as possible to assure an accurate serum level reading.

    (a) 8

    (b) 12

    (c) 16

    (d) 24

ANSWERS

(1) The correct answer is "a," Luvox®/fluvoxamine. It is a SRI, and because it is so serotonergic it is indicated for individuals with obsessive-compulsive disorder. The other three above are SNRIs.

(2) The correct answer is "d," mania and maintenance. The FDA has approved this medication/element for manic episodes of manic-depressive illness and maintenance treatment for manic-depressive patients with a history of mania.

(3) The correct answer is "a," Geodon®/ziprasidone. It is only approved for acute mania and maintenance. The other three are approved for bipolar depression.

(4) The correct answer is "b," 0.6 – 1.2 mmol/L; toxic level is 1.5 – 2 mmol/L; and severe toxicity level is 2.0 mmol/L.

(5) The correct answer is "b," twelve hours.

Module 45

No Pain and More Pain

Opioid Analgesics

😊😊

When it comes to opioids, technically opiates are natural, and opioids are synthetic; however, general parlance refers to the entire panoply of former narcotic treatment options (including natural and synthetic) as opioids. To say that the opioids are a blessing, and a curse is an understatement. For instance, "The physician's painkiller becomes the addict's poison. The poet's dream can be the parent's nightmare. The country's laws are the dealer's opportunity" (Halpern & Blistein, 2019, p. xxii).

In 2017, there were 47,000 opioid related overdoses. In America, one baby born every fifteen minutes is suffering from opioid withdrawal (Halpern & Blistein, 2019). In the United States, in any given year, approximately 100 million people suffer from pain, and 12 million report chronic pain (pain that lasts four months or longer) (Califf, et al., 2016). Yet for every patient suffering intractable cancer, there is blessed relief.

From ancient times, the history of opioids is the same. Pain relief, followed by addiction, followed by management, followed by the enduring possibility of overdose. This cycle plagued Attalus III (170 BCE – 133 BCE) who is historically thought to have performed the first clinical trial to determine doses of various poisons (including opioids) and their antidotes. Nicander of Colophon (197 BCE – 130 BCE), Attalus' research assistant, continued his work and was the first to describe opioid overdose and a proposed antidote. It did not work. This work was continued by Roman physician, Galen (129 CE – 210 CE) who, while studying the effects of opioids on gladiators wounded in battles at the Colosseum, pointed out both the painkilling properties and the courage enhancing properties. He was also responsible for dealing with his patron, Emperor Marcus Aurelius' addiction to opioids. Perhaps the foremost opioid expert in ancient times was Persian physician, Avicenna (980 CE – 1037 CE). He, brilliantly, pointed out that patients seek treatment for two reasons: pain and fear. Opioids relieve both (Halpern & Blistein, 2019; Meldrum, 2016). Closer to today, Thomas Sydenham (1624-1689), introduced "Laudanum" (approximately 10% opioid powder) into medicine, said said, "Of all the remedies it has pleased almighty God to give man to relieve his suffering, none is so universal and so efficacious as opium" (Halpern & Blistein, 2019, p. 175).

Flash forward to more modern times. Four researchers working independently of each other indirectly enhanced the ease of opioid abuse, addiction, and overdose. First, Friedrich Wilhelm Adam Serturner (1783-1841) isolated an "alkaloid," which he named morphine (after Morpheus the Greek god of dreams). Second, Alexander Wood (1817-1884) developed the hypodermic syringe. It is believed that his wife, Rebecca, had a fondness for morphine and may have been the first recorded victim of an overdose after injection. Third, C.R. Alder Wright (1844-1894) developed diacetylmorphine; and finally, Heinrich Dreser (1860-1924) picked up where Wright left off. As a chemist at Baeyer, he took Wright's diacetylmorphine and refined it into "heroiche" (pronounced "here-o-sha") from the German word meaning "heroic" and "strong," and designed this new "heroin" to be a drug for the rich. Dreser fell victim to his creation; he became an addict and likely overdosed in 1924 (Halpern & Blistein, 2019).

The creation of the hypodermic needle, morphine, and heroin contributed to the epidemic of pain relievers, addictions, and overdoses which continue to this day. No doubt the current opioid crisis started in the 19th Century, and with no regulation, was exacerbated by numerous opiate-based patent medicines like "Dover's Powder," "Tonic of Dr. X" and "Miraculous Water of Z."

This need to reduce pain and to alleviate fear was an unspoken "don't talk about it" problem during the Civil War, World War I, and World War II. However, it was the 1980s hospice movement, one that continues to serve so many, that may have unwittingly led to today's opioid difficulties. Enter Purdue Pharma L.P. (founded by John Purdue Gray) and the three infamous, Sackler Brothers (Meier, 2018).

Arthur, Mortimer, and Raymond Sackler purchased Purdue Pharma predominantly for MS Contin, a medication used in palliative and hospice care. Additionally, they bought an advertising agency charged explicitly with branding. This company was tasked with removing the term "narcotic" from the public's lexicon and replacing it with the emotionally neutral term, "opioid."

The next step was to convince physicians that Purdue Pharma's opioids were safe. To achieve this end, a new term, "pseudoaddiction," was coined in 1989. It was a term designed to legitimize a patient's dependency. Roughly defined, pseudoaddiction is a situation in which a physician misunderstands a patient's exhibiting compulsive drug-seeking behavior, one who goes to multiple doctors for scripts for "relief" (today known as "doctor shopping"), and does not recognize that the patient is an opioid dependent. Purdue Pharma went on to argue that the patient is *not* a drug abuser, but one who is suffering from pseudoaddiction and is receiving inadequate medication to manage their pain. According to "pseudoaddiction," the solution was to prescribe more opioids (Meier, 2018; Weissman & Haddox, 1989). Prescribers believed and "bought into" this specious concept; the marketing was a success; and physicians were manipulated.

Because the term, narcotic, was banished to history, the new term, "opioids" became the treatment of choice to manage pseudoaddiction and became the treatment reigning protocol. Narcotics were replaced with a more neutral-sounding name, opioids. As a result, more and more opioids were prescribed. Billions of dollars were made (and continue to be made) on the opioids; millions of people were addicted.

Perhaps the need for opioids is due to increases in real pain, a desire for pain relief, need to escape from the vicissitudes of modern living, or a combination of these. However, there may be an additional reason. In previous decades, somehow, following the false notion of "pseudoaddiction," the concept of "no pain or zero pain" became the prescriber's goal. There is no doubt that this approach has led to overdose deaths by the thousands and addictions by the millions. Zero pain, while a worthy goal, is impossible (Bernard, et al., 2018; Collier, 2018; Meier, 2018; Meldrum, 2016).

The opioids are a cure and a curse. Used correctly, they can be miraculous pain relievers, but they come with a cost. Because they are negatively reinforcing, they are easily abused, and all mental health practitioners must know the pharmacological facts about the opioids.

The receptors associated with the opioids are mu (μ), delta (δ), kappa (κ), and the most recent receptor (discovered in 1995) and whose intrinsic activity has recently been identified is the Nociceptin-Orphanin FQ peptide (NOP; "FQ" is the first and last amino acids of the 17 amino acid peptide, phenylalanine and glutamine). While little is known at this point, it appears this receptor holds great treatment promise.

Of these four receptors, the mu receptor is the most important for current opioid treatment and abuse. Since there is a natural opioid receptor present in the body, the body also makes endogenous neurotransmitters which interact these receptors. The body's natural endogenous pain-relieving neurotransmitters are the endorphins (associated with the mu receptors), the enkephalins (associated with the delta receptors), and the dynorphins (associated with the kappa receptors). As a result, these natural pain-relieving neurotransmitters and their corresponding receptors exert a powerful, life-protective effect. Opioid drugs are exogenous (from the outside) and work on the same receptors in the same way as the body's natural endogenous neurotransmitters.

The mu receptor, agonized by current opioid medications, is found throughout the brain; and it is this receptor which is associated with rapid tolerance, reduction in pain, the feel good, heroic-courageous effect, and negative reinforcement. Concerning overdose, high doses of opioids on the mu receptor of the respiratory center in the rostral ventral medulla inhibit the breathing response triggered by carbon dioxide and result in respiratory depression (defined as a respiration rate less than twelve breaths per minute, failing to ventilate the lungs adequately).

The opioid agonists include morphine, codeine, heroin (diacetylmorphine), Dilauded®/hydromorphone, Oxycontin®/Percodan®/Oxecta®/oxycodone, Demerol®/meperidine, Sublimaze®/Lazanda®/fentanyl citrate nasal spray/fentanyl, Dolophine®/methadone. Antagonists, which are available in many states without a prescription include: Narcan®/Evzio®/naloxone and ReVia®/Vivitrol®/naltrexone. Mixed agonist-antagonists include Buprenex®/Belbuca®/Subutex®/buprenorphine, Suboxone®/buprenorphine and naloxone (combination). Buprenorphine is the prototypic partial opioid agonist.

Opiates have a regular and a reverse tolerance. Tolerance means that it takes more and more of a medication to achieve pain relief. Reverse tolerance means that while pain relief may take higher dosages of a medication, certain other symptoms require less and less to emerge. Constipation is a classic example. While the patient is increasing their dose for pain relief, it takes a lower and lower dose for the constipation side effect to emerge and persist.

The following signs/symptoms are characteristic of chronic use of opioids:

- ↓ Testosterone and ↓ estrogen (Bawor, et al., 2015; Cepeda, et al., 2015; Coluzzi, et al., 2018; Vuong, et al., 2010)
- ↓ Melatonin, interferes with REM and Δ Sleep (Tripathi, et al., 2020)
- ↓ Adenosine (Wu, et al., 2013; Gauthier, et al., 2011)
- ↓ Norepinephrine in the locus coeruleus (Van Bockstaele & Valentino, 2013)
- Biot's Respiration (quick, irregular apnea, named for Camille Biot in 1876) (Van Ryswyk & Antic, 2016)
- Hypertensive rebound (elevation in blood pressure due to withdrawal) (Meyer & Quenzer, 2019)

Stages of increasing intensity are associated with abrupt discontinuation of opioids. "Opioid Withdrawal Syndrome" is categorized by the following stages withcorresponding symptomatology (Atkins, 2018).

- Stage 1: Lasts up to 8 hours; symptoms include anxiety and drug craving.

- Stage 2: From 8 to 24 hours; withdrawal symptoms include anxiety, insomnia, gastrointestinal disturbance, rhinorrhea (runny nose), mydriasis (dilation of pupils), diaphoresis (sweating).

- Stage 3: Up to 3 days; withdrawal symptoms include tachycardia, nausea, vomiting, hypertension, diarrhea, fever, chills, tremors, seizure, and muscle spasms.

Additionally, opioid overdose is characterized by a triad of symptoms. Any emergency medical provider will look for these three symptoms as an indicator of overdose. "The Opioid Triad" includes:

- Pinpoint pupils
- Respiratory depression (less than 12 breaths per minute insufficient to ventilate lungs)
- Coma (decreased level of consciousness)

If there is one thing this current opioid pandemic has demonstrated clearly, drug treatment alone for opioid addiction is not completely effective. Treatment, in addition to medical monitoring, must have other, perhaps equally essential components merged into treatment. These services include family, vocational, mental health, educational, legal, and HIV/AIDS consultations.

It is more than ever a painful reality that "The physician's painkiller becomes the addict's poison. The poet's dream can be the parent's nightmare. The country's laws are the dealer's opportunity" (Halpern & Blistein, 2019, p. xxii). The search for pain relief, desire for "zero pain" management of addictions, and the never-ending parade of funerals from overdoses will never be eradicated, but they can be reduced.

QUIZ TIME

(1) Which of the following is NOT a sign of "The Opioid Triad?"

      (a) pinpoint pupils

      (b) constipation

      (c) respiratory depression

      (d) coma

(2) The first-, second-, and third-generation antipsychotics share one treatment modality in common. It is their ability to treat _____.

      (a) schizophrenia

      (b) personality disorders

      (c) bipolar disorders

      (d) seizure disorders

(3) In addition to medical management, which of the following adjunctive treatment services are considered necessary components in treating opioid addiction?

      (a) legal

      (b) vocational services

      (c) HIV/AIDS

      (d) family

      (e) all the above

(4) Which of the following is NOT an opioid receptor?

      (a) alpha

      (b) mu

      (c) kappa

      (d) delta

(5) Opioids work on the body's receptors, but opioid drugs are considered _____.

      (a) vertiginous

      (b) endogenous

      (c) exogenous

      (d) vertiginous

ANSWERS

(1) The correct answer is "b," constipation. It is a side effect of opioid abuse but not a member of the opioid triad.

(2) The correct answer is "a," schizophrenia.

(3) The correct answer is "e," all the above, along with mental health and educational services.

(4) The correct answer is "a," alpha. Alpha receptors are adrenergic receptors in the sympathetic nervous system.

(5) The correct answer is "c," exogenous, coming from the outside and working on endogenous receptors.

Module 46

Mother's Little Helper

The Benzodiazepines

☺ ☺ ☺

Hard to imagine today, but in the 1950s, office-based mental health treatment was relatively new. Psychoses fell in the domain of hospital treatment. However, with new medications, people began to bring their neuroses, personality disorders, and substance abuse issues into the office setting. Not just psychoanalysis, most of this fell within the psychiatrist's purview. Psychiatry was moving from institutions onto Main Street, and it was the creation of the benzodiazepines which genuinely made it so (Healy, 2002).

The benzodiazepines (BZDs) are psychoactive – they work in the Central Nervous System (brain and spinal cord) in as little as twenty minutes. Their names state their chemical structure, just divide the syllables. These drugs represent a fusion of a "benzene" ring with a "diazepine" ring, hence the name "benzodiazepine." Also, a BZD generic drug typically ends in "lam" or "pam" (e.g., diazepam, alprazolam, midazolam), except for the very first one created in 1955 by Leo Sternbach.

Sternbach's chlordiazepoxide (brand name Librium®) hit the market in 1960, followed by Hoffman-La Roche's Valium®/diazepam. By 1977, after Klonopin®/clonazepam and Ativan®/lorazepam hit the market, the benzodiazepines (affectionately nicknamed "benzos" and a myriad of other street names) were top-selling drugs. Notwithstanding tolerance, dependence, difficult withdrawal, and a litany of side effects, these medications were, and continue to be, a staple of modern psychopharmacology. "Mother's Little Helpers," made famous in a song by The Rolling Stones (specifically referencing Valium®/diazepam; The Rolling Stones, 1966), was preferred as cheaper alternatives to heroin and the more dangerous barbiturates. The BZDs offered exactly what the advertisement slogan of the time promised, "I used to care, but now I take a pill for that."

Whether it is a "lam" or a "pam" or Librium®/chlordiazepoxide, the mechanism of action is the same; the only real difference is each drug's potency and half-life. For instance, 1 mg of Klonopin®/clonazepam and Xanax®/alprazolam is equivalent to 20 mg of Valium®/diazepam. Concerning half-life, Klonopin®/clonazepam has a half-life of 20-50 hours depending on the individual's metabolism. Xanax®/alprazolam runs between 9 and 20 hours. Valium®/diazepam has a half-life of 30 to 60 hours. In other words, "A benzo is a benzo is a benzo."

All BZDs have different potencies and half-lives, but they all have an identical mechanism of action. They are "allosteric modulators of the GABA-A chloride ionophore." They serve as little helpers (allosteric modulators) at their very own, physically distinct, separate binding site on the GABA-A chloride ionophore (fast-acting) receptor. If a BZD binds there without a GABA molecule binding, nothing happens. But if GABA binds at its active site on the GABA-A chloride ionophore and a BZD concomitantly binds at a separate site, it truly is mother's little helper. Relaxation, calm, and peace of mind results.

Old nomenclature did not have an exact abbreviation for BZDs' mechanism of action; however, the reasonably new "Neuroscience-based Nomenclature (NbN)" seems to have gotten it mostly right. NbN refers to BZDs as "GABA-PAM, GABA-Positive Allosteric Modulator." They omit the specific receptor, in this case the GABA-A receptor, where the drug works. There is a GABA-B receptor that works much differently than the GABA-A receptor.

Some courses teach that BZDs are "GABA receptor agonists." That is not completely accurate. Specifically, BZDs only work on the GABA-A chloride ionophore in conjunction with a GABA molecule; BZDs do not work on the GABA-B metabotropic receptor which prefers potassium. GABA-A is ionotropic, and the GABA-B receptor is a 2nd messenger, G protein, metabotropic, cascade-of-action receptor.

Clinical uses for BZDs include: "Anxiolytics for generalized anxiety disorder; panic disorders (including agoraphobia); other phobias; social anxiety; anticonvulsants (especially for alcohol withdrawal); muscle relaxants; and amnesics as an adjunct to anesthetics for procedural memory loss" (Advokat, et al., 2019, p. 502).

As a psychoactive, acute effects provide a rapid change in mood and/or thoughts and are highly addictive. Even though they have a Therapeutic Index (TI) in a safe range, there is a risk for fatal overdose when combined with alcohol, barbiturates, and even some antihistamines. Paradoxically they are calming but can disinhibit (a lack of restraint, impulse control issues, and/or disregard for social conventions along with poor risk assessment and reactions). This disinhibition can affect motor, emotional, cognitive, and perceptual functioning. To an inexperienced diagnostician, this could be misdiagnosed as mania.

Perhaps it is the long-term side effect that is most surprising: Depression. This explains why so many patients are prescribed a BZD for sleep at night and require an antidepressant by day. Emotional blunting, suicidal behavior, and/or psychomotor retardation are common effects. In addition to disinhibition and limited impulse control, memory problems may also be seen with long-term use (Guina & Merrill, 2018).

The risks and limitations taught to every psychopharmacology student, run counter to the actual practice associated with prescribing BZDs. These medications may have decreased utility for anxiety after four months; moreover, researchers suggest that they lose their effectiveness as a sleep aid after three to fourteen days (Guina & Merrill, 2018). Yet, clinicians continue prescribing these medications for extended periods.

Stopping BZDs abruptly is, in a word, dangerous; and the withdrawal reaction is pronounced. The hallmark symptom of withdrawal is textbook depression. Tremor, shakiness, and twitches can be seen along with muscle aches and pains. Because BZDs affect the sleep cycle, withdrawal can create sleep disturbances. A very curious withdrawal side effect is hypersensitivity to noise, light, and touch. The risk of suicide and self-harm, especially in younger people, may present upon withdrawal.

Agonizing the GABA-A chloride ionophore is a treatment or prophylaxis for seizures; abruptly withdrawing from BZDs may generate seizures.

It is also important to note that BZD withdrawal is easily confused by both patients and prescribers as the recurrence of the original anxiety and/or sleep disorder. Abrupt discontinuation of both the BZDs and the SRIs is contraindicated. If the drug treats something, upon withdrawal, the signs/symptoms one was trying to remedy, return stronger and more powerfully. To that point, the BZDs are negatively reinforcing.

Patient education is essential prior to prescribing any BZD and must include the following topics:

- Tolerance and dependence can occur, and dependency associated with dose and duration.
- A time-limited use strategy and an approach to facilitate the successful termination of medication.
- Lethal danger associated with mixing the BZDs with other medications, especially alcohol.
- Abrupt cessation and risk of seizures.
- Additional or alternative treatment modalities, in addition to or instead of medication.

Mother's little helper can help on a time-limited basis. However, too much of mother's little helper fosters addiction and dependence in a short period. Mother's little helper creates tolerance. With the wrong patient and the incorrect diagnosis, mother's little helper is not a helper and can quickly become a burden. Given the wrong prescriptive guidance, the BZDs can rapidly make things worse. Caution advised.

QUIZ TIME

(1) If a generic drug ends in either "lam" or "pam," the chances are good that it is a(n):

    (a) antidepressant

    (b) antipsychotic

    (c) benzodiazepine

    (d) small, cuddly, farm animal

(2) Abrupt withdrawal of the BZDs can cause

    (a) no problem at all

    (b) minimal, slightly uncomfortable withdrawal symptoms

    (c) same symptoms similar to SRI withdrawal

    (d) seizures

(3) BZDs' mechanism of action is a _____.

    (a) BZRA – benzodiazepine receptor agonist

    (b) GABA-PAM – GABA Positive Allosteric Modulator

    (c) GABA-B receptor agonist

    (d) GABA-A receptor agonist

(4) The only real difference between the BZDs is _____ and _____.

    (a) potency and therapeutic index

    (b) therapeutic index and half-life

    (c) half-life and bioavailability

    (d) potency and half-life

(5) Which topics must be covered concerning BZDs and patient education?

    (a) Tolerance can develop quickly

    (b) Other treatment modalities (e.g., therapy) in addition to drug treatment

    (c) Abrupt cessation and corresponding side effects

    (d) Drug reduction and termination strategy

    (e) All the above

ANSWERS

(1) The correct answer is "c," benzodiazepine or BZD.

(2) The correct answer is "d," seizures, along with the possibility of tremor, shakiness, twitches, muscle aches, and pains, interrupted sleep cycle and sleep disturbances, hypersensitivity to noise/light/touch, and the risk of suicide and self-harm (especially in younger people).

(3) The correct answer is "b," GABA Positive Allosteric Modulator. However, a more accurate nomenclature is "allosteric modulator of the GABA-A chloride ionophore." Remember, the new nomenclature, the correct answer listed above, does not delineate the difference between GABA-A and GABA-B receptors.

(4) The correct answer is "d," potency and half-life. All BZDs have the same mechanism of action, but BZDs vary in their potency and half-life.

(5) The correct answer is "e," all the above. These points should be reiterated at each drug review session.

Module 47

Psyched-Up

Amphetamines and Pseudo-amphetamines

☺ ☺

Few drugs trigger the polarity of opinions as stimulants prompt. Some view them as good drugs, even miraculous; other opinions malign the entire class of medications. No doubt many abuse them; and yet stimulants can be life savers given the correct diagnosis. Some people are major proponents of amphetamines as a treatment for ADHD/ADD, yet others question both the diagnosis and the medications. Stimulants will continue to be debated as long as they are prescribed. How these drugs work on the symptoms of these disorders continues to be theory. What they do pharmacologically is not debatable.

There is a difference between "methylphenidate class" stimulant drugs and amphetamines. The difference is in the mechanism of action. As a result, drugs like Ritalin®/methylphenidate (also Concerta®, Metadate® CD, Ritalin® LA, Methylin, QuilliChew® ER, Quillivant® XR, Aptensio® XR, Daytrana®, JORNAY® PM, and Focalin®/methylphenidate-D) are occasionally referred to as "pseudo-amphetamines," "nonamphetamines," and even "non-stimulants." These pseudo-amphetamines create roughly the same response as amphetamines but generate an effect as an indirect agonist. They block the reuptake of the catecholamines, dopamine and norepinephrine. Both amphetamines and pseudo-amphetamines produce sympathomimetic effects (stimulate the sympathetic nervous system) and can provide rapid relief from the symptoms of Attention Deficit Hyperactivity Disorder (ADHD).

Ritalin®/methylphenidate was initially developed in the 1940s from benzyl cyanide by Leandro Panizzon and named for his wife, Rita Lynn. However, some suggest this might be an urban myth and point out its name is derived from its metabolite, a de-esterified urine product called ritalinic acid. It appeared in the United States in the late 1950s (early 1960s according to some sources). It was an overnight success. To this day, given the power of marketing, it is difficult to distinguish which came first, the disorder (ADHD) or Ritalin®/methylphenidate. Some historians believe that both were marketed equally.

In old parlance, Ritalin®/methylphenidate is a "stimulant," blocking the reuptake of norepinephrine and dopamine. The new "Neuro-based Nomenclature (NbN)" now refers to it as a "DN-RIRe," a dopamine norepinephrine reuptake inhibitor and releaser (the NbN-2 app labels it a DN-MM, dopamine norepinephrine multimodal). Its pejorative, slang nickname among clinicians is "Vitamin R." It is unclear why this medication is called a "releaser," as its most well-known mechanism of action is that of a reuptake inhibitor with a slight ability to release dopamine. Frankly, the methylphenidate stimulant class is more like reuptake inhibitors with a cocaine-like action, without the serotonin reuptake blockade. Perhaps for the sake of clarity and to give all stimulants a similar nomenclature, the NbN commission added the "Re," releaser, suffix. This is a clear example that the new NbN nomenclature is a work in progress. One thing is obvious, how the neurotransmitters are blocked by

347

methylphenidate is different from the way amphetamines release them. In comparison, the methylphenidates are more blockers than releasers; and amphetamines are more releasers than blockers. However, the new "Neuro-based Nomenclature" lists both the methylphenidate and amphetamine classes with identical mechanisms of action.

This class of stimulants, methylphenidate, is FDA approved to treat Attention Deficit Hyperactivity Disorder (ADHD in children and adults — approved ages vary based on formulation) and narcolepsy (Metadate® ER, Methylin® ER) (Stahl, 2017).

The amphetamines are far trickier and were developed by Romanian chemist Lazar Edeleanu in 1887 (also credited with creating the procedure to refine crude oil). Calling it "phenylisopropylamine," he noted it had a structure similar to the natural substance ephedrine. Isolated from the herb ma-huang by Nagayoshy Nagai, ephedrine was discovered around the same time (late 19th Century). Time passed, and the actions and effects of amphetamines were not studied until 1910 when British chemist George Barger (also known for his identification of tyramine, the 'mean' monoamine) and Henry Hallett Dale (also known for his research with acetylcholine and shared the Nobel Prize with Otto Loewi) wrote about amphetamine's sympathomimetic effects. However, at that time, it was believed that these discoveries did not have any medicinal, therapeutic effects. No additional study was deemed worthy until 1918/1919 (thirty years after Edeleanu's discovery) when Japanese chemist Akira Ogata simplified the process of synthesizing methamphetamine.

To backtrack briefly, in 1924, ephedrine's structure was discovered to be similar chemically to the neurotransmitter epinephrine. Ephedrine was used to treat asthma; because it was plant based, there was a concern about undersupply. The search was on for a chemical alternative as opposed to a limited plant-based product. There was a need for a working synthetic, ephedrine alternative. As a result, Edeleanu's once-thought worthless amphetamine was resurrected (Advokat, et al., 2019).

Gordon Alles, working in a similar research area, recollected Edeleanu's amphetamines. Alles' new "amphetamine" was a synthetic analog of ephedrine. In the early 1930s, he created an effective decongestant, also appropriate for asthma and sold it under the brand name Benzedrine® (the racemic mixture of amphetamine). Problem solved. However, a new problem was created: Abuse potential.

During World War II, the US, Germany, and Japan gave amphetamines to soldiers to better the fight (i.e., enhance performance) and to treat battle fatigue. In 1935, amphetamines were used to treat narcolepsy; and in 1937, it was acknowledged as a treatment for ADHD. Large-scale abuse emerged in the 1940s, and by the 1960s, amphetamines were abused predominantly for their anorexic and antidepressant actions (Advokat, et al., 2019). Today, amphetamines are FDA approved for the treatment of:

- Attention Deficit Hyperactivity Disorder (ADHD) in children 3-12 (Adderall®, Evekeo®)

- Attention Deficit Hyperactivity Disorder (ADHD) in children ages 6-17 (Adderall® XR, Evekeo®, Dyanavel® XR, Adzenys® XR-ODT)

- Attention Deficit Hyperactivity Disorder (ADHD) and in adults (Adderall® XR, Evekeo®, Adzenys® XR-ODT), Mydayis® (for 13 years and older)

- Narcolepsy (Adderall®, Evekeo®)

- Obesity (Evekeo®)

- Binge eating disorder (Vyvanse®/lisdexamfetamine)

It is important to note the actions of the left and right side of the amphetamine molecule. The left side ("l," "levo," "es") increases norepinephrine in the peripheral nervous system (PNS). The right side ("d" or "dextro") increases dopamine in the central nervous system (CNS) (Berman, et al., 2009; Heal, et al., 2013; Moszczynska, 2016).

Amphetamines are typically taken either orally, IM, or subcutaneous injection (called "skin popping"). They go by many names: "bennies" (an homage to Alles' Benzedrine®), "dexies," "black beauties," "uppers," and "diet pills." Taken orally, a typical 5 to 15 mg dose is absorbed in about thirty minutes, and behavioral effects follow.

Newly defined as both reuptake inhibitors and releasers, it is amphetamine's five "mechanisms of action/release" that is amazing. The steps are as follows:

1. The amphetamine molecule mimics dopamine and "tricks" the dopamine transporter on the terminal button. It does not "sneak in." Because amphetamines are chemically similar to dopamine, it is granted entry, and the amphetamine molecule enters the terminal button.

2. The amphetamine molecule provokes the vesicles to release dopamine into the cytoplasm of the terminal button.

3. The amphetamine molecule reverses the dopamine transporter where it entered, and now dopamine floods the synaptic cleft. The result is a massive increase of dopamine into the extracellular fluid, precipitating a massive increase in dopamine levels in the synaptic cleft.

4. At high doses, amphetamine molecules block monoamine oxidase, and dopamine is not broken down and remains in the cleft longer.

5. Neurotransmitters are released without action potential. The neuron does not depolarize; it does not fire; it releases dopamine into the cleft without the safeguard of neuronal depolarization (Advokat et al., 2019; Meyer & Quenzer, 2019).

The "meth" in methamphetamine indicates a methyl group (CH_3) has been added to the amphetamine molecule. This methyl group makes it more potent, more lipophilic, and easier to cross the blood brain barrier (BBB). More potent, more lipophilic combined with the five mechanisms of action make this once-legal medication, Desoxyn®, a powerful means of abuse. Methamphetamine is five times more likely to produce psychotic behavior. This risk is doubled if used with marijuana or alcohol (McKetin et al., 2013; Moszczynska, 2016).

Stimulant medications are fraught with side effects including:

- COMMON
    - Insomnia
    - Headache
    - Nervousness
    - Irritability
    - Sympathetic overstimulation
    - Tremor
    - Dizziness
    - Abdominal pain
    - Slowed normal growth
    - Blurred vision
    - Increase in tics (movement disorders)
- RARE
    - Psychosis
    - Cardiac problems
    - Seizures
    - Anemia (insufficient healthy red blood cells to carry oxygen)
    - Leukopenia (reduction of white blood cells)

Provigil®/modafinil was discovered accidentally. Laboratoire Louis Lafon, a French pharmaceutical company, searched for a "new" nonsteroidal anti-inflammatory drug (NSAID). While testing a new drug on mice, researchers noticed that one group of rats treated with a new drug became hyperactive. The original drug was "adrafinil," a prodrug that is metabolized into today's modern modafinil (more potent and longer acting than the parent drug). Chemically, Provigil®/modafinil looks more like methylphenidate; however, the mechanism underlying its stimulant activity is more complex. While its exact mechanism is unknown, it binds with weak affinity to the dopamine transporter (DAT) and acts as a weak reuptake inhibitor. While its mechanism of action is important, it is this drug's

"downstream" actions that may be the biggest clue to this medication's power. This action is as follows:

1. Provigil®/modafinil stimulates the release of norepinephrine and orexin. Both neurotransmitters are important in the brain's arousal system.

2. This medication appears to inhibit GABA release.

3. The combined effects of orexin and GABA lead to increased histamine release, a major component of wakefulness. Recall that histamine is essential in every brain for a stream of wakeful consciousness; blocking histamine causes sedation.

4. This medication's collective actions wake you up (histamine) and make you feel great to be awake (dopamine). There is the possibility of abuse. (Meyer & Quenzer, 2019).

Other brand names for Provigil®/modafinil are Alertec® and Modiodal®. This medication has never been officially classified as a stimulant but rather as a "wake-promoting" medication. Its old nomenclature is "DRI" (Dopamine Reuptake Inhibitor), and its new Neuroscience-based Nomenclature (NbN) is the same as the old terminology, "D-RI (Dopamine Reuptake Inhibitor)." It is FDA approved for shiftwork sleep disorder, narcolepsy, and obstructive sleep apnea-hypopnea syndrome (OSAHS — adjunct to standard treatment for underlying airway obstruction) (Stahl, 2017). Provigil®/modafinil is the racemic mixture; Nuvigil®/armodafinil is the right isomer and twice as potent.

While the stimulants are generally regarded as safe, there is "talk" about the degree of neurotoxicity amphetamines express over time. The neurotoxicity of methamphetamine (far more toxic and lipophilic than amphetamines) is known. The following is a list of neurotoxic risks associated with the more potent and lipophilic methamphetamine (Meyer & Quenzer, 2019):

- Hippocampal atrophy (hippocampus is associated with short-term memory)
- Increased neuronal excitotoxicity
- Cardiotoxicity
- Dopamine neuronal death in substantia nigra (indicative of movement disorders)
- Decreased tyrosine hydroxylase (enzyme which converts tyrosine into L-Dopa, dopamine, norepinephrine, and epinephrine)
- Decreased dopamine in dorsal striatum (leads to movement disorders)
- Decreased glucose in brain over time (essential for proper brain functioning)

At the end of this module, the big question becomes, "Are stimulant medications neurotoxic for children who stay on them for years? And what does that neurotoxicity look like?" The answer today is unknown. However, wise clinicians are watchful and remain vigilant for these neurotoxic indicators when prescribing "regular" stimulants.

QUIZ TIME

(1) Amphetamines _____ dopamine; methylphenidate _____ dopamine reuptake.

      (a) release; blocks

      (b) release; releases

      (c) block; releases

      (d) block; blocks

(2) Ritalin's mechanism of action is a(n)_____.

      (a) SNRI

      (b) NDRI

      (c) SSRI

      (d) SNDRI

(3) True or False? Compared to classic amphetamines, methamphetamine is more potent and lipophilic, making it easier to cross the blood brain barrier (BBB).

(4) True or False? Provigil®/modafinil is a wake-promoting medication because it indirectly agonizes histamine and orexin. It is dopamine that is negatively reinforcing.

(5) True or False? When a drug manipulates dopamine – either agonizes or antagonizes – movement disorders can emerge.

ANSWERS

(1) The correct answer is "a," amphetamines release dopamine, and methylphenidate blocks transporter or reuptake.

(2) The correct answer is "b," NDRI, norepinephrine dopamine reuptake inhibitor. SNRI is the mechanism of action for antidepressants like Effexor®/venlafaxine. SSRI explains how antidepressants like Prozac®/fluoxetine work; and SNDRI is the mechanism of action for cocaine.

(3) TRUE. More potent – a little goes a long way. More lipophilic, more fat loving – a little goes a long way. Easier to cross the blood brain barrier – a little goes a long, long way.

(4) TRUE. While it is considered a dopamine reuptake inhibitor, it also works on norepinephrine, orexin, and GABA. Together they manipulate histamine, which is the wakefulness neurotransmitter in the brain.

(5) TRUE. All stimulants, including wake-promoting Provigil®/modafinil, manipulate dopamine. It would not be unusual to see "tics," a movement disorder, emerge as a side effect of these medications. Involuntary sniffing and shrugging of shoulders have been anecdotally reported.

Module 48

One, Two, Three Daily Punch

Nicotine, Caffeine, and Alcohol

☺ ☺ ☺

Former smokers say the best "smoke" of the day is the first one. Probably because the body has slept through four or five half-lives and upon waking is in withdrawal. With nicotine receptors pulsating, the body is now demanding a first dose, quick fix of nicotine. Still moving slowly, an early morning cup of coffee to the rescue. Caffeine is the morning's "wake-up call." Between all-day smoking cigarettes and swilling coffee, to say the body is wired by five in the afternoon is an understatement. Not to worry, pour the body an elixir that acts as a combo dopamine agonist, GABA agonist, glutamate antagonist, and mu-opioid agonist, and the body will chill and settle in for the night. A daily "one-two-three punch" of nicotine, followed by a second punch of caffeine, concluded with a third punch of alcohol.

Of the legal drugs, it is believed that nicotine, caffeine, and alcohol are the most widely abused medications in the world. Of course, that was before cannabis (i.e., marijuana). Because cannabis is legal in some states and not others, nicotine, caffeine, and alcohol continue to make up the big three. However, there is little doubt that cannabis is rapidly becoming the most used/abused drug in the world. However, it is not likely nicotine, caffeine, and alcohol consumption will lose favor because of cannabis' popularity. More in the following module.

Nicotine. This addictive substance was named after Jean Nicot de Villemain, a French diplomat, who was the first to introduce tobacco (i.e., snuff) to Europe. Nicotine is a drug with no current therapeutic applications in medicine. With toxicities well known, the most addictive substance known to humans, it is the leading cause of preventable death and disease in the United States. Although more than 4,000 compounds are released via a burning cigarette, nicotine is the primary addictive substance. From the tobacco plant, nicotine is a natural insecticide secreted by the plant to fend off insects. However, it does not repel humans; they smoke it. The average cigarette yields ten puffs within five minutes, and the typical individual absorbs 1 to 2 milligrams (absorption can range from 0.5 to 3 milligrams). An average smoker absorbs 20 to 40 milligrams daily (Advokat, et al., 2019).

Nicotine agonizes the nicotinic receptors (named after "nicotine," the first substance discovered to stimulate this receptor). The receptors are abbreviated, "nAChRs" (nicotinic acetylcholine receptors). Nicotine has biphasic action on receptors. First it stimulates, and then it blocks. Desensitization follows and can occur after just one cigarette. Now the receptor will temporarily not respond to either nicotine or acetylcholine. Additionally, other chemicals in cigarette smoking inhibit monoamine oxidase A and B, which cause an increase in dopamine and likely account for the "feel good" that comes with cigarette smoking.

There is one positive nicotinic note. Nicotine is "nootropic" (from Greek, "noos" meaning "mind," and "trope" meaning "turning"). Coined in 1972 by Corneliu E. Giurgea, nootropics describe a future class of medications that act selectively to enhance memory and mind. As every college student from earlier decades knows, smoking while studying makes it easier to remember. In other words, nicotine is a memory enhancer.

Acute effects of nicotine include blood pressure and heartbeat increase and a release of epinephrine from the adrenal glands (adrenal medulla). First-time smokers notice nausea and vomiting because nicotine stimulates the brain's chemical trigger zone (CTZ, vomiting center). With limited tolerance to nicotine, nicotine serves as a laxative and causes fluid retention by stimulating the hypothalamus to release antidiuretic hormone. There is also a decrease in muscle tone which likely explains the "relaxation" that follows smoking a cigarette (Advokat, et al., 2019). Chronic use of nicotine/tobacco may result in cancer, cardiovascular disease, and respiratory diseases (emphysema and bronchitis).

Caffeine. Most people start their day with a cup of "stimulating" coffee and rarely acknowledge or realize that it is imbued with one of the world's most "abused" drugs. Originally discovered in Ethiopia around 850 CE, it arrived in Europe in 1615, and the rest is history. In 1819, it was isolated by 25-year-old Friedlieb Ferdinand Runge. There was a brief period when caffeine was thought to be hazardous. For this reason, legal charges were brought against Coca-Cola in 1909. The legal case, *United States vs Forty Barrels and Twenty Kegs of Coca-Cola* (1916), dragged on for years, and Coca-Cola finally agreed to reduce the amount of caffeine (cocaine had been removed years earlier in 1895, although the exact amount of cocaine originally in Coca-Cola is unknown). In 1916, legal action was dropped (Sepkowitz, 2013). This court case is alleged to have disclosed, in open court, the seven secret ingredients (known as "7X") in Coca-Cola.

It is believed that 85% of the United States population (excluding children under two years of age) has at least one caffeinated beverage per day (Mitchell, 2014; Ferré, 2013). Dietary caffeine is in beverages (coffee, tea, soft drinks), foods, drugs, and now a separate category including "energy drinks." The amount of caffeine varies wildly in traditional beverages. Gourmet coffee can average 200 milligrams per 8 fluid ounces. Make that a grande, and 16 ounces equals 400 milligrams of caffeine or more. The FDA guidelines (2008) state that 400 milligrams per day of caffeine consumption are consumer-safe for adults (except for mothers-to-be) and half that amount for adolescents (Advokat, et al., 2019). There is evidence that intake of 250 milligrams or more may interfere with sleep (Meyer & Quenzer, 2019).

Caffeine powder is unregulated and may be purchased at health food stores, tobacco shops, and on-line. While it resembles "sugar," a spoonful can quickly lead to overdose. For instance, 1/16th tsp of caffeine powder equals approximately 250 milligrams or the amount of caffeine in three Red Bull® energy drinks. One teaspoon of caffeine powder equals the caffeine in 50 Red Bull® energy drinks. The lethal dose is 10 grams, although three grams ingested in a short period of time can be fatal.

When taken orally, caffeine is completely and rapidly absorbed into the system and peaks in 30 minutes. Its mechanism of action is an adenosine antagonist at A_1 receptors and weakly at A_2 receptors (capital letter "A," stands for adenosine). Adenosine is a wakefulness neurotransmitter and builds up throughout the day and encourages sleep. Caffeine blocks these receptors and tricks the brain into thinking it is more awake than it really is. In this regard, it is a stimulant, but with a different mechanism of action from classic amphetamines, pseudo-amphetamines, and cocaine, yet all are members of the class "stimulants." While the half-life varies from person to person (from 2 hours to 10 hours), the average half-life (t½) of caffeine is easy to remember courtesy of the energy drink "Five-Hour ENERGY®." Remember this energy drink's name, and the t½, five hours, is easy to remember.

Regular use of caffeine leads to tolerance and dependence. Excessive caffeine use can also lead to two disorders: (1) caffeine intoxication and (2) caffeine dependence syndrome or caffeine use disorder. Caffeine intoxication (too much, too fast to the point of poisoning the system) is characterized by restlessness, nervousness, insomnia and physiological problems like tachycardia, myoclonus (muscle twitching), and gastrointestinal distress. Caffeine dependence is related to chronic, maladaptive caffeine use. Adverse effects (occurring with heavy consumption, 12 or more cups of coffee per day) from "caffeinism" (clinical syndrome produced by overuse) include enhanced anxiety, increased agitation, tachycardia, hypertension, cardiac arrhythmias, and gastrointestinal distress.

Withdrawal effects are unpleasant, notably a first-rate headache along with fatigue, drowsiness, sad mood (including depression), irritability, inability to concentrate, flu-like symptoms (including nausea and vomiting), and muscle pain/stiffness. Upon a headache's arrival, people usually ingest caffeine to resolve the withdrawal symptoms. In this way, caffeine is negatively reinforcing.

There are a few health benefits. First, caffeine is used to treat sleep apnea in newborn babies. It also has a variety of positive effects on both mental and physiological health. Other benefits include heightened attention, reduced fatigue, wakefulness, improved sports/psychomotor performance, increased memory consolidation, and pain relief (Meyer & Quenzer, 2019). Finally, caffeine causes sperm to swim faster and more vigorously toward their target (Burhman & Phelan, 2000).

Alcohol. As early as 8000 BCE, the first alcoholic beverage, mead, was brewed from fermented honey. Today, alcohol remains a central nervous system depressant, creates a feel good, relaxed, worry-free, and pain-free mental state. Unlike most drugs cleared by first-order kinetics, alcohol is cleared by zero-order kinetics. Roughly one ounce of alcohol is cleared per hour independent of the volume consumed. No matter how much alcohol is imbibed, the rate of clearance remains unchanged. Binge drinking, a pattern of drinking that brings a person's blood alcohol concentration (BAC) to 0.08 g/dl (grams per deciliter), is defined as five drinks for men and four for women in two hours (Meyer & Quenzer, 2019).

Withdrawal after heavy use can last for days. Symptoms are tremor, anxiety, high blood pressure, high heart rate, diaphoresis (sweating), rapid breathing, nausea, and vomiting. Severe withdrawal may involve "delirium tremens" and consists of hallucinations, convulsions, disorientation, and painful anxiety.

Because alcohol is easily absorbed and distributed, it affects most of the body's organ systems. However, given individual different body types and drinking styles, the exact point where alcohol becomes damaging is not clearly defined. Long-term alcohol abuse leads to Vitamin B_1 deficiency which causes cell death in the periaqueductal gray (PAQ), thalamus, and mammillary bodies. Wernicke's encephalopathy (tremors, weakness, ataxia, confusion, and disorientation) may follow. With continued abuse, Korsakoff's syndrome follows with permanent damage to thalamic nuclei, leading to enduring memory loss with confabulation. Over time, liver damage includes fatty liver, changes to alcohol-induced hepatitis, and concludes with cirrhosis (Meyer & Quenzer, 2019).

Alcohol has a diuretic effect (need to urinate). It increases sexual desire but inhibits performance. Alcohol is also teratogenic (causes damage to an unborn fetus). Prenatal exposure may produce "Alcohol-Related Neurodevelopmental Disorder" (previously called "Fetal Alcohol Syndrome"). This is characterized by intellectual disabilities, developmental delays, low birth weight, and craniofacial malformations.

The behavioral effects of alcohol are causally related to the patient's blood alcohol level and are listed below.

BLOOD ALCOHOL LEVEL AND PHYSIOLOGICAL RESPONSE

Blood Alcohol Level (BAC)	Physiological Response
0.04%	Measurable effects commence
0.08%	Legal definition of "Driving Under Influence" (varies by state)
.15 — .18	Major impairment, mental and physical Slurred speech Exaggerated emotions Serious loss of judgment Large increase in reaction time
.15	Vomiting starts
.20 — .25	Staggering
.30	Conscious but stuporous
.35	Loss of consciousness
.40	Lethal Dose 50 (LD_{50})
.50	Toxic Death

One area that is not often discussed is the "medicinal aspects" of alcohol. Alcohol use at mild to moderate levels may have these effects:

- Alcohol Decreases:
  - Proinflammatory cytokines (McCarter, et al., 2017)
  - Gallstones (Cha, et al., 2019; Wang, et al., 2017)
  - Kidney stones (Littlejohns, et al., 2020)
  - Diabetes (Holst, C., et al., 2017)
  - Heart disease (Chiva-Blanch & Badimon, 2019)
  - Platelet aggregation (Di Castelnuovo, et al., 2010)
  - LDL ("bad" cholesterol) (Advokat, et al., 2019)
- Alcohol Increases:
  - HDL (Healthy lipoproteins) (Advokat, et al., 2019)
  - Vasodilation (Advokat, et al., 2019)

To summarize, (1) Nicotine stimulates and then relaxes. (2) Caffeine stimulates. (3) Alcohol depresses/relaxes and turns down the brightness of the day. It is almost a perfect trifecta, a "run of three daily wins." Use one to get started; use the other to stimulate throughout the day; then use another to calm down after all the stimulation. They are powerful. They are and will be abused forever – or until something better and cheaper comes along. Standby for cannabis.

QUIZ TIME

(1) The FDA advises that _____ milligrams, daily intake of caffeine, is regarded as safe.

      (a) 200

      (b) 300

      (c) 400

      (d) 500

(2) The average cigarette smoker daily absorbs _____ of nicotine.

      (a) 10 to 20 mg

      (b) 20 to 40 mg

      (c) 30 to 40 mg

      (d) 40 to 50 mg

(3) Cigarette smoking biphasically agonizes and then blocks the _____ receptor.

      (a) nicotinic-acetylcholine receptors (nAChr)

      (b) dopamine

      (c) serotonin

      (d) histamine

(4) Binge drinking is defined as _____ drinks for women and _____ drinks for men in a two-hour period.

      (a) 1 & 2

      (b) 3 & 4

      (c) 4 & 5

      (d) 4 & 4

(5) The sequence of liver deterioration due to or caused by prolonged alcohol abuse is _____.

      (a) cirrhosis to alcohol hepatitis to fatty liver disease

      (b) fatty liver disease to alcohol hepatitis to cirrhosis

      (c) alcohol hepatitis to fatty liver disease to cirrhosis

      (d) alcohol hepatitis to cirrhosis to fatty liver disease

ANSWERS

(1) The correct answer is "c," 400 mg daily for the average adult. It is approximately half that for children. However, safety factors/dosages of caffeine intake for pregnant women are not established (Advokat, et al., 2019).

(2) The correct answer is "b," 20-40 mg per day, however this amount can vary.

(3) The correct answer is "a," nicotinic-acetylcholine receptors.

(4) The correct answer is "c," four drinks for women and five for men within a two-hour time frame.

(5) The correct answer is "b," fatty liver disease turns to alcohol hepatitis which can turn to cirrhosis.

Module 49

Baked

The Cannabinoids

☺☺

Many readers are tempted to start this module first. After all, cannabis is everywhere; allegedly "everyone" uses it; numerous places are selling it, including dark alleys and street corners. And physical conditions that respond to its medicinal properties are being identified each day along with an increase in the number of people with government-granted cards that allow them to purchase it. It is now big business, really big business. As the timbre of the time becomes more and more complicated, as the exigencies and vicissitudes of life become more and more unchecked and unpredictable, the use of caffeine, alcohol, and nicotine will pale compared to the increased use of cannabis, marijuana. Already marijuana is the most heavily used illicit drug in the United States (Meyer & Quenzer, 2019). Its status from illegal to legal will likely change.

If it were not for Canada and Israel (and some states where cannabis is legal), the world would know little of this plant's medicinal qualities. While cannabis may be the earliest known plant not cultivated for food, history is replete with stories of this plant's being used for exhilaration and medication. Cannabis was brought to the United States by Spanish explorers around the mid-16th Century. Cultivated for rope in naval use, and since it did not grow well in England's clime, in 1611 Sir Walter Raleigh was ordered to cultivate hemp in Virginia (alongside tobacco). In the British colonies, it was a staple for rope, clothes, and paper. Once Napoleon's army brought it back from Egypt in the 19th Century, it spread for medical and recreational purposes.

Questions surround the exact meaning of the word "cannabis." Some sources suggest it means "reed or cane-like." Other elder sources say it came from the Hebrew term, "kaneh-bosem," meaning "fragrant cane," as referenced in Exodus 30:23. The ancient Egyptians, long before the pyramids, called cannabis, "shemshemet." Ancient Chinese gave it its own logogram symbol which looks like a hemp plant growing upside down. The name, cannabis, is thought to have started its life as a Scythian or Thracian term until it migrated and was "borrowed" by the Romans and Greeks. Finally, in 1753, Carl Linnaeus gave this weed-like, flowering plant a botanical name: *Cannabis sativa* (from Latin, "sativum," meaning "cultivated").

It was marketed in America as a "tincture of cannabis," commonly used until the Pure Food and Drug Act (1906) and Harrison Narcotic Act (1914) imposed restrictions. In the 1930s, Harry J. Anslinger (first commissioner of the Bureau of Narcotics) effectively maligned cannabis through media attacks and legislation. Because cannabis is negatively reinforcing, Anslinger's attempts would in time fail socially. A substantial part of the population never accepted his attempt to disparage cannabis. As a result, marijuana found a forever place in parts of society.

It is currently a Schedule I drug – meaning it is alleged to have no medicinal value. While policies are relaxing, its current inconsistent legal status continues to produce problems from state to state. According to the National Institute on Drug Abuse (NIDA, 2020), marijuana is the most commonly used psychotropic drug in the United States after alcohol. More prevalent among men than women, and widespread among adolescents and young adults.

Cannabis is unique; it has its own classification system; its mechanism of action is novel. Many psychopharmacology students when asked, "What is cannabis' mechanism of action?" are taught to respond, "It hits everything." While that cannot possibly be entirely true, it is close. While there is still much to be learned about all that cannabis does pharmacodynamically, here is a partial list:

- Decreases adenylate cyclase (catalyzes ATP's conversion to cAMP and is a regulator of stored energy). This accounts for cannabis' "laid back" or "baked" effect (Bryant, et al., 2018; Condie, et al., 1996).

- Decreases glutamate in the hippocampus. Glutamate plays a role in long-term memory consolidation. This explains why a person may believe they have "world-changing illuminations" while "enjoying" the drug but cannot remember the idea when the drug wears off. (Coluzzi, et al., 2019; Fan, et al., 2010; Scarante, et al., 2017).

- Decreases ACh, DA, NE, and 5-HT. By reducing acetylcholine at neuromuscular junctions, movement is reduced. Dopamine production is decreased, and this may account for the "time slowing down" perception while using the drug. Reduced dopamine in the spiny neurons in the striatum creates the perception of time crawling. Reduction in norepinephrine reduces energy. Reduced serotonin will also slow things down, produce irritability, and may partially explain the food cravings, "the munchies" (Meyer & Quenzer, 2019).

- Decreases Ca^{2+}. Calcium plays a role in neurotransmitter release from the terminal button (exocytosis). This may explain the decreases in monoamine release and continue to facilitate the "baked feeling" (Ross, et al., 2008; Scarante, et al., 2017).

- AChEI in CNS (?). This is a questionable area and is currently being researched. By blocking acetylcholinesterase (AChE) in the brain, it increases acetylcholine in the brain and may reduce dementia symptoms (Eubanks, et al., 2006).

- Facilitates potassium (K^+). CB_1 (cannabinoid number 1) receptors activate potassium channel openings (Meyer & Quenzer, 2019). Because potassium affects heart muscles, increased potassium can be dangerous. It is known that marijuana use can increase heart rate for up to three hours after use (Meyer & Quenzer, 2019).

- Mu (μ) opiate agonist. Stimulating this receptor reduces pain perception (Befort, 2015; Maguire & France, 2014).

- Stimulates dopamine firing in the ventral tegmental area (VTA) and releases dopamine in the nucleus accumbens, the "reward center." As a result, cannabis is self-reinforcing, and the probability of "doing it again" is increased (Bloomfield, et al., 2016; Martz, et al., 2016).

- More and more "what cannabis does to the body" will be reported as research is "allowed."

Chronic cannabis use has been shown to have harmful effects including:

- Respiratory problems (Meyer & Quenzer, 2019)
- Increased risk of myocardial infarction (Meyer & Quenzer, 2019)
- Suppression of immune functions (Meyer & Quenzer, 2019)
- Adverse effects on offspring development when used by pregnant mothers (Meyer & Quenzer, 2019)
- Interferes with male and female reproductive system
  - Decreases testosterone (Payne, et al., 2019)
  - Decreases sperm count (Payne, et al., 2019)
  - Decreases sperm viability (percentage of live sperm in a semen sample) (Payne, et al., 2019)
  - Decreases Follicle Stimulating Hormone (FSH, stimulates growth of ovarian follicles before the release of an egg from one follicle; also increases estradiol production) (Dunne, 2019).
  - Decreases Luteinizing Hormones (LH, controls the function of ovaries in females and testes in men) (Dunne, 2019).
  - Menstrual cycles altered (Brents, 2016; Walker, et al., 2019)

The body has its own cannabinoid (note the word "cannabis" is the root of the word "cannabinoid") receptors, demonstrating that the body makes natural neurotransmitters that fit this receptor. In other words, the body makes its "own natural weed" neurotransmitters. There are two known natural, endogenous (body made) cannabinoid (eCB) neurotransmitters discovered by Lumir Hanus and William Devane in 1992. The first, anandamide (also known as N-arachidonoylethanolamide, AEA), is derived from the Sanskrit word, "ananda," meaning "bringer of bliss, joy, and delight." It is derived from eicosatetraenoic acid, an essential omega-6 fatty acid.

The other eCB is 2-arachidonoylglycerol (2-AG) and presents at larger levels in the brain than anandamide. Unlike "The Big Seven" neurotransmitters, endogenous cannabinoid neurotransmitters are made on demand, only when needed. Only after the neuron is stimulated will it produce anandamide and/or 2-AG. Another difference is that the cannabinoid neurotransmitters are retrograde signalers. They are released from postsynaptic cells, diffuse into the cleft, and stimulate presynaptically, not postsynaptically. Why they are retrograde is not clearly understood. One theory is that they are neuroprotective and can terminate terminal button action should it release too many neurotransmitters.

Two cannabinoid receptors, CB_1 and CB_2 have been discovered. CB_1 is the main cannabinoid receptor in the brain (notably in the basal ganglia, cerebellum, hippocampus, and cerebral cortex). When agonized, it has been shown to impair learning, memory consolidation, and short-term memory.

CB_2, first located in the immune system, is also expressed in microglial cells which remove waste, eliminate damaged neurons, and fight infections. They are essential for maintaining a healthy central nervous system (CNS). Agonizing the CB_2 receptor stimulates cytokine release and immune cell movement toward an inflammatory site. These two receptors (CB_1 and CB_2) are metabotropic, G protein-coupled, second messenger receptors, and involve a cascade of action.

Exogenous compounds (coming from outside via the cannabis plant) work on these same CB_1 and CB_2 receptors. The euphoric effects of smoking marijuana appear to be mediated by the CB_1 receptor. This receptor is also involved in the brain's reward system, increasing the probability of repeated use and/or abuse will follow.

Important to the field of psychopharmacology, the cannabis plant contains over 70 unique compounds. These are collectively known as the "phytocannabinoids." This term simply means that these compounds have a cannabinoid shape. Phytocannabinoids are plant-derived cannabinoids that interact with the endocannabinoid system (ECS).

The psychoactive phytocannabinoid, Δ^9-tetrahydrocannabinol (THC) is likely responsible for cannabis' recreational use. There is also a non-psychoactive component, cannabidiol (CBD). Important for its emerging potential, this constituent (CBD) may be appropriate for treating various health and mental health issues. There is an inverse relationship between the psychoactive component, THC, and the non-psychoactive component, CBD. CBD keeps THC in check. As CBD increases, the psychoactive effects of THC decrease (Advokat, et al., 2019).

An individual's age of use significantly affects the biological process. Regular users below seventeen years of age have smaller whole brain volumes, less cortical gray matter, yet greater amounts of cerebral blood flow. Additionally, THC increases "cognitive effort" for simple inhibitory tasks. It slows processing speed, increases memory problems, and decreases the ability to attend.

The elimination half-life is 20 to 30 hours, but gradual movement of cannabis' fat-soluble metabolites can be detected in the human system for two weeks or longer. This fact holds even after one, single marijuana use. Withdrawal symptoms emerge one day after discontinuation, peak at about one week, and can last up to three weeks. Withdrawal symptoms include: Cravings, anorexia, irritability, restlessness, and sleep disturbances.

Cannabis is purported to treat numerous disorders; however, more rigorous testing to authenticate treatment is needed. The following is a partial list:

- Nausea
- Appetite
- Analgesia (inflammatory and neuropathic pain)
- Decrease interocular pressure (IOC)
- Bronchodilation
- Seizures (increase seizure threshold, making it more difficult to seize)
- Tourette Syndrome
- Multiple sclerosis
- Menstrual difficulties

Since the early 2000s, synthetic designer cannabinoids have been marketed. "K2," "Spice," and numerous others are highly potent, full agonists of both CB_1 and CB_2 receptors. They are quite capable of producing severe states of intoxication. Adverse health effects can include kidney damage, seizures, panic attacks, psychosis, and possible fatalities. The designer cannabinoids are significantly dangerous substances.

Even with all the possible adverse effects, cannabis continues to be a valued commodity that has once again found a prominent place in modern society. The plant's metabolites are masterful in their ability to help individuals feel good, forget troubles, and remove pain. It has been around for thousands of years, and it will continue to be around for thousands more. It is too reinforcing to ever go away.

QUIZ TIME

(1) Which of the following endocannabinoid receptors is related to an individual's immunity?

       (a) CB_1

       (b) CB_2

       (c) CB_3

       (d) CB_1 and CB_2

(2) Cannabis can be detected in a urine test up to _____ after use.

       (a) one week

       (b) two weeks

       (c) three weeks

       (d) four weeks

(3) Which of the following cannabinoid metabolites is thought to hold the most interest for medical therapeutics?

       (a) CBD

       (b) THC

       (c) Δ^9

       (d) anandamide

(4) Endocannabinoid receptors are _____.

       (a) postsynaptic

       (b) ionotropic

       (c) metabotropic

       (d) retrograde signalers

(5) True or False? Endocannabinoids are held in vesicles and released from the terminal button into the synaptic cleft.

ANSWERS

(1) The correct answer is "b," the CB_2 receptor.

(2) The correct answer is "b," two weeks. Cannabis can be sequestered in a person's fat. This is also dependent on the frequency of use.

(3) The correct answer is "a," CBD, the cannabidiols.

(4) The correct answer is "c," metabotropic, second-messenger, G protein-coupled, cascade of action receptors.

(5) FALSE. Endocannabinoids are only made "when requested." They diffuse into the cleft, and then, utilizing retrograde signaling, bind with presynaptic CB_1 or CB_2 receptors. They are broken down by Fatty Acid Amide Hydrolase (FAAH) and Monoacylglycerol Lipase (MAGL).

Module 50

Here's to the Future

Pain, Memory, Top Ten Side Effects

☺

In some ways, this last unit is the most demanding, hardest to achieve module. This module represents hope that the future will find "the perfect drug." One that works. One with a simple mechanism of action that is easy to understand. One that cures. One that has no side effects. One that works for everyone who takes it. And, perhaps most important of all, one that is inexpensive and affordable to all.

It is believed that the next twenty years will be two decades associated with the mantra, "Treat pain and memory, first." As a result, the treatment of mental illness may take a back seat. As baby boomers age and the "iGeneration" continues to demand immediate fixes, the expectation that pain is "fixed with a pill" will persist.

Modern society does not understand that "pain" functions as a signal something is wrong. It is a "request" that something needs to be fixed. Currently, society seems content to "give me something that makes the pain go away." Many psychotropics mitigate pain. The old SNRIs ("NbN" SN-RIs) lessen pain by about 30%, but it does not treat the source (Ghaemi, 2019). Then, there are the opioids. They work well for acute pain but cause problems in the long run as a treatment for chronic pain. However, the negative reinforcing aspects of these medications suggest that the opioids are not going anywhere. Supply and demand apply to medical treatments, too.

Treating memory is trickier. At present, there are no successful remedies for memory disorders. Aricept®/donepezil is FDA approved to treat Alzheimer's disease (mild, moderate, and severe). It works by inhibiting acetylcholine's breakdown, thereby indirectly agonizing and increasing acetylcholine throughout the entire body. Its mechanism of action is an acetylcholinesterase inhibitor (AChEI) or given the new NbN nomenclature, ACh-EI, acetylcholine enzyme inhibitor. In this case, the old terminology is better because it defines explicitly the enzyme, acetylcholinesterase, that is being inhibited or blocked.

The problem with medications like Aricept®/donepezil is its lack of receptor specificity. By increasing acetylcholine all over the body, even though the prescriber only wants the medication to work on the brain's memory center, a multitude of side effects ensue. Recall that acetylcholine is integral to the workings of the parasympathetic nervous system, sympathetic nervous system, somatic system, and sympathetic innervation of the adrenal medulla. The side effects of Aricept®/donepezil are numerous and not limited to "SLUDGEM" – salivation, lacrimation, urination, defecation, GI distress, emesis, and miosis. No wonder numerous patients frequently discontinue this medication (approximately eight percent or one in every twelve).

Razadyne®/galantamine works similarly to Aricept®/donepezil, but with more multimodal action. "Neuroscience-based Nomenclature (NbN)" defines it as a ACh-MM (acetylcholine multimodal, an enzyme inhibitor, positive allosteric modulator). Translated: Acetylcholinesterase inhibitor with nicotinic receptors boosted. However, there is one "memory helper drug" that is slightly different.

Exelon®/rivastigmine works like Aricept®/donepezil, the same mechanism of action (acetylcholinesterase inhibitor), old and NbN new (ACh-EI), but with a twist. It, too, increases acetylcholine all over the body including the memory spot in the brain. However, this drug also works by inhibiting butyrylcholinesterase, a non-specific cholinesterase enzyme. As a result, this dual action (AChEI and BuChEI) could enhance therapeutic efficacy (Anderson, et al., 2019). Additionally, Exelon®/rivastigmine is broken down at the site of action. It does, however, present with "SLUDGEM" side effects. In fact, the butyrylcholinesterase boost may cause more GI distress. Like other memory medications in this class, it may slow the progression, but it does not, regrettably, reverse cognitive impairment.

Finally, and concluding back where this working textbook started with the ultimate desire to protect the patient, there are ten life-threatening side effects of psychotropic medications. All providers involved in integrated care from the receptionist to the chief of medicine can spot them – that is, if one knows what to look for.

The rule is as follows: It may not be your responsibility to treat side effects, but it is all mental health professionals' responsibility to report them immediately to the prescriber so they may assess if it is, in fact, a life-threatening side effect. For instance, a rash could be nothing more than poison ivy, or it could be Stevens-Johnson

Syndrome. Be safe. Refer to the prescriber. If it is life threatening, the patient is spared a difficult recovery. Another good example is a sore throat. It could be a cold, or maybe the medication has a side effect called "agranulocytosis" (immunity compromised). Be safe. Refer it to the prescriber. Once again, the patient is protected.

Ask the patient at the beginning of every session, "Have you noticed any physical changes since our last meeting?" Listen for their answer. Match it to the "Top Ten Life-Threatening Side Effects." If their physical complaint is on the list, refer to the prescriber and document action. It is not recommended to show the patient the entire list. Simply ask, "Have you noticed any physical changes since our last meeting?" Catching any one of these ten life-threatening side effects protects the patient.

TOP TEN LIFE-THREATENING PSYCHOTROPIC SIDE EFFECTS
(1) Allergic reaction, rashes
(2) Change in level of alertness, consciousness, racing thoughts
(3) Eating problems
(4) Changes in bowel function
(5) Changes in heartbeat
(6) Fainting or dizziness
(7) Abnormal posture, movement, gait
(8) Yellowing of eyes, skin, or mouth
(9) Unusual bruising or bleeding
(10) Sudden cold or sore throat

QUIZ TIME

(1) Serotonin-norepinephrine reuptake inhibitors mitigate pain by approximately
_____.

     (a) 10%

     (b) 20%

     (c) 30%

     (d) 40%

(2) Opioids work best for _____ pain.

     (a) acute

     (b) chronic

     (c) pseudoaddiction

     (d) both a & b

(3) The current mechanism of action of today's "memory medications" blocks which enzyme?

     (a) monoamine oxidase (MAO)

     (b) catechol-o-methyltransferase (COMT)

     (c) Fatty Acid Amide Hydrolase (FAAH)

     (d) acetylcholinesterase (AChE)

(4) Approximately _____ of dementia patients experience "SLUDGEM" so severely they discontinue their medication.

     (a) 8%

     (b) 15%

     (c) 18.5%

     (d) 20%

(5) A knowledge of psychopharmacology helps a clinician _____.

     (a) better protect their patients

     (b) provide proper care for patients

     (c) enhance all treatment options

     (d) improve interactions with all level of providers

     (e) all the above

ANSWERS

(1) The correct answer is "c." SNRIs can mitigate the perception of pain by as much as 30%.

(2) The correct answer is "a." Opioids work better for acute pain than chronic pain.

(3) The correct answer is "d." By blocking acetylcholinesterase (AChE), acetylcholine increases in the basal forebrain which is thought to be associated with memory.

(4) The correct answer is "a." Approximately 8% or one in twelve patients will discontinue their dementia treatment due to SLUDGEM – salivation, lacrimation, urination, defecation, GI distress, emesis, and miosis.

(5) The correct answer is "e," all the above. A knowledge base of mental health treatments that includes psychopharmacological tenets protects the patient, provides an added level of care, allows the clinician to better decide the right course of treatment, and improves interactions within all level of mental health providers. It also allows the clinician to answer questions about how medications work, their side effects, and potential risks and benefits.

In conclusion, it is "easy" to prescribe medications. All it takes is a pen and a prescription pad. Getting it right the first time is the trick. Determining the correct diagnosis, mentally juggling all drugs and patient issues, and selecting the proper application (medication, psychotherapy, or both) is a skill that requires knowledge and expertise. It further necessitates a willingness to continually monitor and evaluate what is best for the patient, no matter who raises the question. Indeed, proficient mental health professionals welcome questions rather than fearing them. They see drug queries as an aid to treatment and not a challenge of skill level, prescribing integrity, or self-esteem.

Proficiency in this field requires continual study throughout one's clinical career. Completing these fifty-programmed modules represents a significant step toward your desire for mastery and patient safety.

REFERENCES

Adams, M. P., Holland, Jr., L. N., & Bostwick, P. M. (2010). *Pharmacology for nurses: A pathophysiologic approach.* Upper Saddle River, NJ: Pearson-Prentice Hall.

Ads that can make you sick. (2009, October). *Consumer Reports on Health*, p. 1+.

Adverse drug events are on the rise. (2012). *Consumer Reports on Health, 24*, 2+.

Advokat, C. D., Comaty, J. E., & Julien, R. M. (2014). *Julien's primer of drug action* (13th ed., 40th Anniversary Edition). New York: Worth Publishing.

Advokat, C. D., Comaty, J. E., & Julien, R. M. (2019). *Julien's primer of drug action* (14th ed.). New York: Worth Publishing.

Ahlquist, R. P. (1948). A study of the adrenotropic receptors. *American Journal of Physiology, 153*, 586-600.

Ahmed, A. O., & Bhat, I. A. (2014). Psychopharmacological treatment of neurocognitive deficits in people with schizophrenia: A review of old and new targets. *CNS Drugs, 28*(4), 301-318.

Allighieri, D. (1314). *Inferno*.

American Psychological Association (APA). (2002). Criteria for evaluating treatment guidelines. *American Psychologist, 57*, 1052-1059.

American Psychological Association (APA). (2011). Practice guidelines regarding psychologists' involvement in pharmacological issues. *American Psychologist, 66*, 835-849.

Anderberg, R. H., Richard, J. E., Eerola, K., Lopez-Ferreras, L., Banke, E., Hansson, C., Nissbrandt, H., Berqquist, F., Gribble, F. M., Reimann, F., Wernstedt Asterholm, I., Lamy, C. M., & Skibicka, K. P. (2017). Glucagon-like peptide 1 and its analogs act in the dorsal raphe and modulate central serotonin to reduce appetite and body weight. *Diabetes, 66*(4), 1062-1073.

Andersson, M. L., Moller, A. M., & Wildgaard, K. (2019). Butyrylcholinesterase deficiency and its clinical importance in anaesthesia: A systematic review. *Anaesthesia, 74*(4), 518-528.

Anthony, W., Rogers, E. S., & Farkas, M. (2003). Research on evidence-based practices: Future directions in an era of recovery. *Community Mental Health Journal, 39*, 101-114.

APA Presidential Task Force on Evidence-Based Practice. (2006). Evidence-based practice in psychology. *American Psychologist, 61*, 271-285.

Armani, F., Andersen, M., & Galduróz, J. (2014). Tamoxifen use for the management of mania: A review of current preclinical evidence. *Psychopharmacology, 231*(4), 639-649.

Armstrong, D. (2003, November 18). Study finds increase in medication errors at US hospitals. *The Wall Street Journal*.

Arnold, M. (2008). Polypharmacy and older adults: A role for psychology and psychologists. *Professional Psychology: Research & Practice, 39*, 283-289.

Arthur, G. (2015). Epinephrine: A short history. *The Lancet Respiratory Medicine, 3*, 350-351.

Atkins, C. (2018). *Opioid use disorders*. Eau Claire, WI: PESI Publishing & Media.

Axelrod, J., Whitby, L., & Hertting, G. (1961). Effect of psychotropic drugs on the uptake of H3-norepinephrine by tissues. *Science, 133*, 383-384.

Bachmann, C. J., Roessner, V., Glaeske, G., & Hoffmann, F. (2015). Trends in psychopharmacologic treatment of tic disorders in children and adolescents in Germany. *European Child & Adolescent Psychiatry, 24*(2), 199-207.

Bahar, M. A., Setiawan, D., Hak, E., & Wilffert, B. (2017). Pharmacogenetics of drug-drug interaction and drug-drug dene interaction: A systematic review on CYP2C9, CYP2C19 and CYP2D6. *Pharmacogenomics, 18*, 701-739.

Baldwin, D. S., Anderson, I. M., Nutt, D. J., Allgulander, C., Bandelow, B., den Boer, J. A., Christmas, D. M., Davies, S., Fineberg, N., Lidbetter, N., Malizia, A., McCrone, P., Nabarro, D., O'Neill, C., Scott, J., van der Wee, N., & Wittchen, H. U. (2014). Evidence-based pharmacological treatment of anxiety disorders, post-traumatic stress disorder and obsessive-compulsive disorder: A revision of the 2005 guidelines from the British Association for Psychopharmacology. *Journal of Psychopharmacology, 28*(5), 403-439.

Barber, C. (2008, February/March). The medicated Americans. *Scientific American Mind, 19,* 45+.

Barber, T. R., Griffanti L., Muhammed, K., Drew, D. S., Bradley, K. M., McGowan, D. R., Crabbe, M., Lo, C., Mackay, C. E., Husain, M., Hu, M. T., & Klein, J. C. (2018). Apathy in rapid eye movement sleep behaviour disorder is associated with serotonin depletion in the dorsal raphe nucleus. *Brain: A Journal of Neurology, 141*, 2848-2854.

Barlow, D. H., et al. (1998). Psychosocial treatments for panic disorders, phobias, and generalized anxiety disorders. In P. E. Nathan. & J. M. Gorman (Eds.), *A guide to treatments that work.* New York: Oxford University Press.

Barlow, D. H., Gorman, J. M., Shear, M. K., & Woods, S. W. (2000). Cognitive-behavioral therapy, imipramine, or their combination for panic disorder: A randomized controlled trial. *Journal of American Medical Association, 283*, 2529-2536.

Barondes, S. H. (2003). *Better than Prozac: Creating the next generation of psychiatric drugs.* New York: Oxford University Press.

Barry, P. (2010, January/February). Drug prices up inflation down. *AARP Bulletin, 51*, 10+.

Bateman, A. & Fonagy, P. (2008). 8-year follow-up of patients treated for borderline personality disorder: Mentalization-based treatment versus treatment as usual. *American Journal of Psychiatry, 165*, 631-638.

Bateman, J., Gilvarry, E., Tziggili, M., Crome, I. B., Mirza, K., & McArdle, P. (2014). Psychopharmacological treatment of young people with substance dependence: A survey of prescribing practices in England. *Child and Adolescent Mental Health, 19*(2), 102-109.

Bawor, M., Bami, H., Dennis, B. B., Plater, C., Worster, A., Varenbut, M., Daiter, J., Marsh, D. C., Steiner, M., Anglin, R., Coote, M., Pare, G., Thabane, L., & Samaan, Z. (2015). Testosterone suppression in opioid users: A systematic review and meta-analysis. *Drug and Alcohol Dependence, 149*, 1-9.

Beck, K. (2019). What ending is typically found at the ends of enzymes? *Sciencing.* https://sciencing.com/different-types-enzymes-4968363.html.

Beecher, H. K. (1955). The powerful placebo. *Journal of American Medical Association, 159*, 1602-1606.

Beckett, A. H., & Casy, A. F. (1954). Synthetic analgesics: Stereochemical considerations. *The Journal of Pharmacy and Pharmacology, 6*(12), 986–1001.

Beckett, A. H., & Casy, A. F. (1965). Analgesics and their antagonists: Biochemical aspects and structure-activity relationships. *Progress in Medicinal Chemistry, 4*, 171–218.

Befort, K. (2015). Interactions of the opioid and cannabinoid systems in reward: Insights from knockout studies. *Frontiers in Pharmacology, 6*, 6.

Begley, S. (2010, February 8). The depressing news about antidepressants. *Newsweek*, 35+.

Belar, C. D. (2000). Scientist-practitioner not equal to science plus practice: Boulder is bolder. *American Psychologist, 55*, 249-250.

Benjamin, D. M. (2003). Reducing medication errors and increasing patient safety: Case studies in clinical pharmacology. *Journal of Clinical Pharmacology, 43*, 768-783.

Berlin, R. K., Butler, P. M., & Perloff, M. D. (2015). Gabapentin Therapy in Psychiatric Disorders: A Systematic Review. *The primary care companion for CNS disorders, 17*(5), 10.4088/PCC.15r01821. https://doi.org/10.4088/PCC.15r01821

Berman, S. M., Kuczenski, R., McCracken, J. T., & London, E. D. (2009). Potential adverse effects of amphetamine treatment on brain and behavior: A review. *Molecular Psychiatry, 14*, 123-142.

Bernard, S. A., Chelminski, P. R., Ives, T. J., & Ranapurwala, S. I. (2018). Management of pain in the United States-A brief history and implications for the opioid epidemic. *Health Services Insights, 11*, 1-6.

Bernardy, N. C., & Friedman, M. J. (2015). Psychopharmacological strategies in the management of posttraumatic stress disorder (PTSD): What have we learned? *Current Psychiatry Reports, 17*(4), 564.

Bero, L., & Rennie, D. (1995). The Cochrane Collaboration: Preparing, maintaining, and disseminating systematic reviews of the effects of health care. *Journal of American Medical Association, 274*, 1935-1938.

Beyond the prescription bottle. (2008, Spring). Smart choices. *AARP Bulletin, 49*, 6.

Bezerra, L.S., Santos-Veloso, M.A.O., Bezerra Junior, N.D.S., Fonseca, L.C.D., & Sales, W.L.A. (2018). Impact of cytochrome P450 2D6 (CYP2D6) genetic polymorphism in tamoxifen therapy for breast cancer. *Brazilian Journal of Gynecology and Obstetrics, 40*, 794-799.

Bjorklund, A., & Dunnett, S. B. (2007). Fifty years of dopamine research. *Trends in Neurosciences, 30,* 185-187.

Blair D. T., & Dauner, A. (1992). Extrapyramidal symptoms are serious side-effects of antipsychotic and other drugs. *Nurse Practitioner, 17,* 56, 62-64.

Bloomfield, M. A., Ashok, A. H., Volkow, N. D., & Howes, O. D. (2016). The effects of Delta-9-tetrahydrocannabinol on the dopamine system. *Nature, 539,* 369-377.

Blum, N., St. John, D., Pfohl, B., Stuart, S., McCormick, B., Allen, J., Arndt, S., & Black, D. W. (2008). Systems training for emotional predictability and problem solving (STEPPS) for outpatients with borderline personality disorder: A randomized controlled trial and 1-year follow-up. *American Journal of Psychiatry,* 165, 468-478.

Bodine, W. K. (2007, April). E-prescribing can make a difference for Medicare Part-D. *Pharmacy Times,* 73, 101-103.

Bogenschutz, M. P., Forcehimes, A. A., Pommy, J. A., Wilcox, C. E., Barbosa, P., & Strassman, R. J. (2015). Psilocybin-assisted treatment for alcohol dependence: A proof-of-concept study. *Journal of Psychopharmacology, 29*(3), 289-299.

Bolea-Alamañac, B., Nutt, D. J., Adamou, M., Asherson, P., Bazire, S., Coghill, D., Heal, D., Müller, U., Nash, J., Santosh, P., Sayal, K., Sonuga-Barke, E., & Young, S. J. (2014). Evidence-based guidelines for the pharmacological management of attention deficit hyperactivity disorder: Update on recommendations from the British Association for Psychopharmacology. *Journal of Psychopharmacology, 28*(3), 179-203.

Botulinum toxin. (2017, January 18). *Time,* p. 38.

Brain facts, A primer on the brain and nervous system. (2008). Washington, D.C.: Society for Neuroscience.

Brents, L. K. (2016). Marijuana, the endocannabinoid system and the female reproductive system. *The Yale Journal of Biology and Medicine, 89*(2), 175–191.

Brown, J., & Minami, T. Outcomes-informed care: An evidence-based meta-method for improving outcomes in behavioral healthcare. In *Center for Clinical Informatics Transforming Data into Information.* http://www.clinical-informatics.com/Outcomes%20Informed%20Care%20FAQ.htm

Brownlee, S. (2012, April 23-30). The doctor will see you – If you're quick. *Newsweek,* 46-50.

Brunton, L. L., et al. (2010). *Goodman & Gilman's the pharmacological basis of therapeutics.* New York: McGraw Hill.

Bryant, L. M., Daniels, K. E., Cognetti, D. M., Tassone, P., Luginbuhl, A. J., & Cury, J. M. (2018). Therapeutic cannabis and endocannabinoid signaling system modulator use in otolaryngology patients. *Laryngoscope Investigative Otolaryngology, 3,* 169-177.

Buchanan, R. B., Kreyenbuhl, J., Kelly, D. L., Noel, J. M., Boggs, D. L., Fischer, B. A., Himelhoch, S., Fang, B., Peterson, E., Aquino, P. R., & Keller, W. (2010). The 2009 schizophrenia PORT psychopharmacological treatment recommendations and summary statements. *Schizophrenia Bulletin, 36,* 71-93.

Buck v. Bell, 274 U.S. 200, 47 S. Ct 584; L. Ed. 1000; 1927 U.S. LEXIS 20.

Buffington, P. W. (2012). Creating a culture of medication safety. *The Florida Psychologist, 63*(2), 9+.

Bullmore, E. (2018). *The inflamed mind*. London: Short Books.

Buoli, M., Cumerlato-Melter, C., Caldiroli, A., & Altamura, A. C. (2015). Are antidepressants equally effective in the long-term treatment of major depressive disorder? *Human Psychopharmacology: Clinical & Experimental, 30*(1), 21-27.

Burnham, T., & Phelan, J. (2000). *Mean Genes*. Cambridge, MA: Perseus Publishing.

Bussing, R., Reid, A. M., McNamara, J. P., Meyer, J. M., Guzick, A. G., Mason, D. M., Storch, E. A., & Murphy, T. K. (2015). A pilot study of actigraphy as an objective measure of SSRI activation symptoms: Results from a randomized placebo controlled psychopharmacological treatment study. *Psychiatry Research, 225*(3), 440-445.

Butcher, J. N., Holey, J. M., & Mineka, S. (2014). *Abnormal psychology*. Boston: Pearson

Caccavale, J. L., & Wiggins, J. G. (2009). Adverse drug events: Implications for prescribing psychotropic medications. American Board of Behavioral Healthcare Practice, http://abbhp.org/ade.pdf.

Cade, J. F. J. (1949). Lithium salts in the treatment of psychotic excitement. *The Medical Journal of Australia, 2*, 518-520.

Califf, R. M., Woodcock, J. M., & Ostroff, S. (2016). A proactive response to prescription opioid abuse. *New England Journal of Medicine, 374*, 1480-1485.

Cantor, D. S., & Evans, J. R. (Eds.). (2013). *Clinical neurotherapy: Application of techniques for treatment* (pp. 55-84). San Diego, CA, US: Elsevier Academic Press.

Carbon, M., Hsieh, C. H., Kane, J. M., & Correll, C. U. (2017). Tardive dyskinesia prevalence in the period of second-generation antipsychotic use: A meta-analysis. *Journal of Clinical Psychiatry, 78*, 264-278.

Carlsson, A. (1959). The occurrence, distribution, and physiological role of catecholamines in the nervous system. *Pharmacology Review, 11*, 490-493.

Carpenter, D., Zucker, E. J., & Avorn, J. (2008). Drug-review deadlines and safety problems. *New England Journal of Medicine, 358*, 1354-1361.

Carrell, A. (1935). *Man, the unknown*. New York: Harper and Brothers, 318.

Castren, E., & Rantamaki, T. (2010). The role of BDNF and its receptors in depression and antidepressant drug action: Reactivation of developmental plasticity. *Developmental Neurobiology, 70*, 289-297.

Cepeda, M. S., Zhu, V., Vorsanger, G., & Eichenbaum, G. (2015). Effect of opioids on testosterone levels: Cross-sectional study using NHANES. *Pain Medicine, 16*, 2235-2242.

Cha, B. H., Jang, M. J., & Lee, S. H. (2019). Alcohol consumption can reduce the risk of gallstone disease: A systematic review with a dose-response meta-analysis of case-control and cohort studies. *Gut and Liver, 13*, 114-131.

Chambless, D. L., et al. (1996). An update on empirically validated therapies. *The Clinical Psychologist, 49*, 5-18.

Children's adverse events. (2008, April 7). Reported on ABC "Good Morning America."

Chilvers, C., Dewey, M., Fielding, K., Gretton, V., Miller, P., Palmer, B., Weller, D., Churchill, R., Williams, I., Bedi, N., Duggan, C., Lee, A., & Harrison, G. (2001). Antidepressant drugs and generic counselling for treatment of major depression in primary care: Randomized trial with patient preference arms. *British Medical Journal, 322*, 772-775.

Chiva-Blanch, G., & Badimon, L. (2019). Benefits and risks of moderate alcohol consumption on cardiovascular disease: Current findings and controversies. *Nutrients, 12*, 108.

Choi, M. R., Kouyoumdzian, N. M., Rukavina Mikusic, N. L., Kravetz, M. C., Roson, M. I., Fermepin, M. R., & Fernandez, B. E. (2015). Renal dopaminergic system: Pathophysiological implications and clinical perspectives. *World Journal of Nephrology, 4*, 196-212.

Chorpita, B. F. (2003). The frontier of evidenced-based practice. In A. E. Kazdin & J. R. Weisz (Eds.), *Evidenced-based psychotherapies for children and adolescents* (pp. 42-59). New York: Guilford.

Christodoulou, C., & Kalaitzi, C. (2005). Antipsychotic drug-induced acute laryngeal dystonia: two case reports and a mini review. *Journal of Psychopharmacology, 19*(3), 307–311.

Chwalisz, K. (2003). Evidence-based practice: A framework for the twenty-first-century scientist-practitioner training. *The Counseling Psychologist, 31*, 497-528.

Clark, D. M., Layard, R., Smithies, R., Richards, D. A., Suckling, R., & Wright, B. (2009). Improving access to psychological therapy: Initial evaluation of two UK demonstration sites. *Behaviour Research and Therapy, 47*, 910-920.

Coffey, E. C., & Brumback, R. A. (2006). *Pediatric neuropsychiatry*. Philadelphia, PA: Lippincott Williams & Wilkins.

Cohen, M. R. (1999). *Medication errors*. Washington, D.C.: American Pharmacists Association.

Collier, R. (2018). A short history of pain management. *CMAJ : Canadian Medical Association Journal, 190*(1), E26–E27.

Coluzzi, F., Billeci, D., Maggi, M., & Corona, G. (2018). Testosterone deficiency in non-cancer opioid-treated patients. *Journal of Endocrinological Investigation, 41*, 1377-1388.

Condie, R., Herring, A., Koh, W. S., Lee, M., & Kaminski, N. E. (1996). Cannabinoid inhibition of adenylate cyclase-mediated signal transduction and interleukin 2 (IL-2) expression in the murine T-cell line. EL 4.IL-2. *The Journal of Biological Chemistry, 271*, 13175-13183.

Connor, D., & Banga, A. (2014). A model of pediatric psychopharmacology-psychotherapy treatment integration in the ambulatory clinical setting. *Journal of Child & Family Studies, 23*(4), 686-703.

Cooney, S. M., Huser, M., Small, S., & O'Connor, C. (2007). Evidence-based programs: An overview. What Works, Wisconsin, Research to Practice Services, 1-8.

Cooper, J. K., Love, D. W., & Raffoul, P. R. (1982). Intentional prescription nonadherence (noncompliance) by the elderly. *Journal of American Geriatrics Society, 30*, 329-333.

Correa, J., Heckman, B., Marquinez, N., Drobes, D., Unrod, M., Roetzheim, R., & Brandon, T. (2014). Perceived medication assignment during a placebo-controlled laboratory study of varenicline: Temporal associations of treatment expectancies with smoking-related outcomes. *Psychopharmacology, 231*(13), 2559-2566.

Crits-Cristoph, P. (1998). Psychosocial treatments for personality disorders. In P. E. Nathan & J. M. Gorman (Eds.). *A guide to treatments that work*. New York: Oxford University Press.

Croen, L. A., Grether, J. K., Yoshida, C. K., Odouli, R., & Hendrick, V. (2011). Antidepressant use during pregnancy and childhood autism spectrum disorders. *Archives of General Psychiatry, 68*, 1104-1112.

Crystal, S., Sambamoorthi, U., Walkup, J. T. & Akincigil, A. (2003). Diagnosis and treatment of depression in the elderly Medicare population: Predictors, disparities, and trends. *Journal of American Geriatrics Society, 51*, 1718-1728.

Davis, M. C., Horan, W. P., & Marder, S. R. (2014). Psychopharmacology of the negative symptoms: Current status and prospects for progress. *European Neuropsychopharmacology, 24*(5), 788-799.

de Craen, A.J.M., Kaptchuk, T.L., Tijssen, J.G.P., & Kleijnen, J. (1999). Placebo and placebo effects in medicine: Historical overview. *Journal of the Royal Society of Medicine, 92*, 511-515.

de Vos, A., van der Weide, J., & Loovers, H. M. (2011). Association between CYP2C19*17 and metabolism of amitriptyline, citalopram and clomipramine in Dutch hospitalized patients. *The pharmacogenomics journal, 11*, 359–367.

Davis, T. C., Federman, A. D., Bass, P. F., 3rd, Jackson, R. H., Middlebrooks, M., Parker, R. M., & Wolf, M. S. (2009). Improving patient understanding of prescription drug label instructions. *Journal of General Internal Medicine, 24*, 57-62.

Denic, A., Glassock, R. J., & Rule, A. D. (2016). Structural and functional changes with the aging kidney. *Advances in Chronic Kidney Disease, 23*, 19-28.

Department of Clinical Epidemiology and Biostatistics, McMaster University Health Sciences Centre. (1981). How to read clinical journals. *Canadian Medical Journal, 124*, 555-590.

Depression Guideline Panel. Depression in Primary Care; Treatment of Major Depression: Clinical Practice Guideline. US Department of Health and Human Services, Public health service, Agency for Health Care Policy and Research. AHCPR Publication 93-0551, Rockville, MD.

Depression in primary care. (1993). AHCPR pub 93-0550. Rockville, MD: Agency for Health Care Policies.

Di Castelnuovo, A., Costanzo, S., Donati, M. B., Iacoviello, L., & de Gaetano, G. (2010). Prevention of cardiovascular risk by moderate alcohol consumption: Epidemiologic evidence and plausible mechanisms. *Internal and emergency medicine, 5*, 291-297.

Diller, L. H. (2006). *The last normal child*. Westport, CT: Praeger.

Dixon, L. B., Dickerson, F., Bellack, A. S., Bennett, M., Dickinson, D., Goldberg, R. W., Lehman, A., Tenhula, W. N., Calmes, C., Pasillas, R. M., Peer, J., & Kreyenbuhl, J. (2010). The 2009 schizophrenia PORT psychosocial treatment recommendations and summary statements. *Schizophrenia Bulletin, 36*, 48-70.

Dr. Earl W. Sutherland Jr. Dies; 1971 Nobel Laureate in Medicine. (1974, March 10). *The New York Times*, p. 59. Retrieved from https://www.nytimes.com/1974/03/10/archives/dr-earl-w-sunderland-jr-dies1971-nobel-laureate-in-medicine.

Drake, R. E., Goldman, H. E., Leff, H. S., Lehman, A. F., Dixon, L., Mueser, K. T., & Torrey, W. C. (2001). Implementing evidence-based practices in routine mental health service settings. *Psychiatric Services, 52*, 179-182.

Drake, R. E., Merrens, M. T., & Lynde, D. W. (2005). *Evidence-based mental health practice*. New York: W.W. Norton & Company.

Drugs that can affect your sex life. *Consumer Reports on Health, 24*, 6.

Dunne, C. (2019). The effects of cannabis on female and male reproduction: More high-quality evidence is needed before physicians can reassure patients that marijuana use will not affect their fertility or their offspring. *British Columbia Medical Journal, 61*(7), 282–285.

Dziegielewski, S. F. (2006). *Psychopharmacology handbook for the non-medically trained*. New York: W.W. Norton & Company.

Eadala, P., Waud, J. P., Matthews, S. B., Green, J. T., & Campbell, A. K. (2009). Quantifying the 'hidden' lactose in drugs used for the treatment of gastrointestinal conditions. *Alimentary Pharmacology & Therapeutics, 29*, 677-687.

Eban, K. (2019). *Bottle of lies*. New York: HarperCollins.

Elkin, I., Shea, M. T., Watkins, J. T., Imber, S. D., Sotsky, S. M., Collins, J. F., Glass, D. R., Pilkonis, P. A., Leber, W. R., Docherty, J. P., Fiester, S. J., & Parloff, M. B. (1989). National Institute of Mental Health Treatment of Depression Collaborative Research Program: General effectiveness of treatments. *Archives of General Psychiatry, 46*, 971-982.

El-Mallakh, R. S., Gao, Y., & Roberts, R. J. (2011a). Tardive dysphoria: The role of long-term antidepressant use in-inducing chronic depression. *Medical Hypotheses, 76*, 769-773.

El-Mallakh, R. S., Gao, Y., Briscoe, B. T., & Roberts, R. J. (2011b). Antidepressant-induced tardive dysphoria. *Psychotherapy & Psychosomatics, 80*, 57-59.

El-Merahbi, R., Loffler, M., Mayer, A., & Sumara, G. (2015). The roles of peripheral serotonin in metabolic homeostasis. *Federation of European Biochemical Societies Letters, 589*, 1728-1734.

Emanuel, E. J. (2020). *Which country has the world's best health care?* New York: Public Affairs.

E-prescribing, what it means for you, Part D and the future of written prescriptions. (2008, Spring). Smart choices, *AARP Bulletin, 49*, 2+.

Essock, S. M., Schooler, N. R., Stroup, T. S., McEvoy, J. P., Rojas, I., Jackson, C., & Covell, N. H. (2011). Effectiveness of switching from antipsychotic polypharmacy to monotherapy. *American Journal of Psychiatry, 168*, 702-708.

Estabrook, R. W. (2003). A passion for P450s (remembrances of the early history of research on cytochrome P450). *Drug Metabolism and Disposition: The Biological Fate of Chemicals, 31*, 1461-1473.

Eubanks, L. M., Rogers, C. J., Beuscher, A. E., Koob, G. F., Olson, A. J., Dickerson, T. J., & Janda, K. D. (2006). A molecular link between the active component of marijuana and Alzheimer's disease pathology. *Molecular Pharmaceutics, 3*, 773-777.

Everett, G.M. & Toman, J.E.P. (1959). Mode of action of rauwolfia alkaloids and motor activity. In J. Masserman (Ed.), *Biological Psychiatry*: The Proceedings of the scientific sessions of the society of biological psychiatry (pp. 75-81). New York: Grune & Stratton.

Excessive psychotropic medication & psychotropic medication side effects. Health & Safety Alert #11-03-08. http://test.mr.state.oh.us/health/alerts/alert-11-3-08.pdf.

Factor, S. A., Burkhard, P. R., Caroff, S., Friedman, J. H., Marras, C., Tinazzi, M., & Comelia, C. L. (2019). Recent developments in drug-induced movement disorders: A mixed picture. *Lancet Neurology, 18*, 880-890.

Fairburn, C. G., Cooper, Z., & Shafran, R. (2003). Cognitive behavior therapy for eating disorders: A "transdiagnostic" theory and treatment. *Behaviour Research and Therapy, 41(5)*, 509-528.

Fan, N., Yang, H., Zhang, J., & Chen, C. (2010). Reduced expression of glutamate receptors and phosphorylation of CREB are responsible for in vivo Delta-9-THC exposure-impaired hippocampal synaptic plasticity. *Journal of Neurochemistry, 112*, 691-702.

Fantasia. (1940). Burbank, CA: The Walt Disney Company.

Ferré, S. (2013). Caffeine and Substance Use Disorders. *Journal of Caffeine Research, 3*, 57-58.

Fick, D. M., Cooper, J. W., Wade, W. E., Waller, J. L., Maclean, J. R., & Beers, M. H. (2003). Updating the Beers criteria for potentially inappropriate medication use in older adults. *Archives of Internal Medicine, 163*, 2716-2724.

Field, T. S., Gilman, B. H., Subramanian, S., Fuller, J. C., Bates, D. W., & Gurwitz, J. H. (2005). The costs associated with adverse drug events among older adults in the ambulatory setting. *Medical Care, 42*, 1171-1176.

Fincham, J. E. (2005). *Taking your medicine: A guide to medication regimens and compliance for patients and caregivers*. New York: Pharmaceutical Products Press.

Fincham, J. E., & Wertheimer, A. I. (1985). Using the health belief model to predict initial drug therapy defaulting. *Social Science and Medicine, 20*(1), 101-105.

Flint, R. B., Roofthooft, D. W., van Rongen, A., van Lingen, R. A., van den Anker, J. N., van Dijk, M., Allegaert, K., Tibboel, D., Knibbe, C. A. J., & Simons, S. H. P. (2017). Exposure to acetaminophen and all its metabolites upon 10, 15, and 20mg/kg intravenous acetaminophen in very-preterm infants. *Pediatric Research, 82*, 678–684.

391

Foltz, R. (2011). Re-ED principles in evidence-based standards. *Reclaiming Children & Youth, 19*, 28-32.

Food and Drug Administration. (2018, 12 12). *Spilling the Beans: How Much Caffeine is Too Much?* Retrieved from Food and Drug Administration: https://www.fda.gov/consumers/consumer-updates/spilling-beans-how-much-caffeine-too-much.

Fournier, J. C., DeRubeis, R. J., Hollon, S. D., Dimidjian, S., Amsterdam, J. D., Shelton, R. C., & Fawcett, J. (2010). Antidepressant drug effects and depression severity: A patient-level meta-analysis. *Journal of American Medical Association, 303*(1), 47-53.

Frick, P. J. (2007). Providing the evidence for evidenced-based practices. *Journal of Clinical Child and Adolescent Psychology, 36*, 2-7.

Friedman, M. J., & Davidson, J. T. (2014). Pharmacotherapy for PTSD. In M. J. Friedman, T. M. Keane, & P. A. Resick (Eds.), *Handbook of PTSD: Science and practice (2nd ed.)* (pp. 482-501). New York, NY, US: Guilford Press.

Friedrich, J. O., Adhikari, N., Herridge, M. S., & Beyene, J. (2005). Meta-analysis: Low-dose dopamine increases urine output but does not prevent renal dysfunction or death. *Annals of Internal Medicine, 142*, 510-524.

From neurobiology to treatment: Bipolar disorder and schizophrenia unraveled. (2009). Carlsbad, CA: Neuroscience Education Institute.

Furukawa, T. A., Watanabe, N., & Churchill, R. (2007). Combined psychotherapy plus antidepressants for panic disorder with or without agoraphobia. *Cochrane Database Syst Rev*, CD004364.

Gabbard, G. O., Gunderson, J. G., & Fonagy, P. (2002). The place of psychoanalytic treatments within psychiatry. *Archives of General Psychiatry, 59*, 505+.

Gagnon, M. A., & Lexchin, J. (2008). The cost of pushing pills: A new estimate of pharmaceutical promotion expenditures in the United States. *PLoS Medicine, 8*, 1-5.

Gallagher-Thompson, D., & Steffen, A. M. (1994). Comparative effects of cognitive-behavioral and brief psychodynamic psychotherapies for depressed family caregivers. *Journal of Consulting and Clinical Psychology, 62*, 543-549.

Gauthier, E. A., Guzick, S. E., Brummett, C. M., Baghdoyan, H. A., & Lydic, R. (2011). Buprenorphine disrupts sleep and decreases adenosine concentrations in sleep-regulating brain regions of Sprague Dawley rat. *Anesthesiology, 115*, 743-753.

Geisen-Bloo, J., van Dyck, R., Spinhoven, P., van Tilburg, W., Dirksen, C., van Asselt, T., Kremers, I., Nadort, M., & Arntz, A. (2006). Outpatient psychotherapy for borderline personality disorder: Randomized trial of schema-focused therapy vs. transference-focused psychotherapy. *Archives of General Psychiatry, 63*, 649-658.

Genetic Information Non-Discrimination Act. US Department of Health and Human Services. www.hhs.gov/hipaa/for-professionals/special-topics/genetic-information/index.html.

Getz, G. E. (2014). *Applied biological psychology.* New York: Springer Publishing Company.

Ghaemi, S. N. (2019). *Clinical psychopharmacology: Principles and practice*. New York: Oxford University Press.

Gilbert, D. A., Altshuler, K. Z., Rago, W. V., Shon, S. P., Crismon, M. L., Toprac, M. G., & Rush, A. J. (1998). Texas medication algorithm project: Definitions, rationale, and methods to develop medication algorithms. *Journal of Clinical Psychiatry, 59*, 345-351.

Glenmullen, J. (2005). *The antidepressant solution*. New York: Free Press.

Goisman, R. M., Warshaw, M. G., & Keller, M. B. (1999). Psychosocial treatment prescriptions for generalized anxiety disorder, panic disorder, and social phobia, 1991-1996. *American Journal of Psychiatry, 156*, 1819-1821.

Goldberg, J. F., & Ernst, C. L. (2019). *Managing the side effects of psychotropic medications*. Washington, D.C.: American Psychiatric Association.

Goldstein, D. (Ed.). (2015). *Oxford companion to sugars and sweets*. Oxford, England: Oxford University Press.

Goodman, K. W. (2002). *Ethics and evidence-based medicine*. Cambridge: Cambridge University Press.

Gordon Guyatt presentation at the XV Cochrane colloquium in Sao Paulo, Brazil. http://www.colloquiumbrasil.info/program/public/documents/plenary_1b_guyatt-151144.htm

Green, W. H. (2007). *Child and adolescent clinical psychopharmacology*. New York: Lippincott Williams and Wilkins.

Grilly, D. M., & Salamone, J. D. (2012). *Drugs, brain, and behavior*. New York: Pearson.

Grunze, H., Erfurth, A., Marcuse, A., Amann, B., Normann, C., & Walden, J. (1999). Tiagabine appears not to be efficacious in the treatment of acute mania. *The Journal of Clinical Psychiatry, 60*, 759-762.

Guina, J., & Merrill, B. (2018). Benzodiazepines I: Upping the care on downers: The evidence of risks, benefits and alternatives. *Journal of Clinical Medicine, 7*, 17+.

Gunduz, G. U., Parmak Yenner, N., Kilincel, O., & Gunduz, C. (2018). Effects of selective serotonin reuptake inhibitors on intraocular pressure and anterior segment parameters in open angle eyes. *Cutaneous and Ocular Toxicology, 37*, 36-40.

Hacker, J. (2008). *The great risk shift: The new economic insecurity and the decline of the American dream*. New York: Oxford University Press.

Hallak, J. E. C., Dursun, S. M., Bosi, D. C., de Macedo L. R. H., Machado-de-Sousa, J. P., Abrao, J., Crippa, J. A. S., McGuire, P., Krystal, J. H., Baker, G. B., & Zuardi, A. W. (2011). The interplay of cannabinoid and NMDA glutamate receptor systems in humans: Preliminary evidence of interactive effects of cannabidiol and ketamine in healthy human subjects. *Progress in Neuro-Psychopharmacology & Biological Psychiatry, 35*, 198-202.

Halpern, J. H., & Blistein, D. (2019). *Opium*. New York: Hachette Books.

Hantsoo, L., Ward-O'Brien, D., Czarkowski, K., Gueorguieva, R., Price, L., & Epperson, C. (2014). A randomized, placebo-controlled, double-blind trial of sertraline for postpartum depression. *Psychopharmacology, 231*(5), 939-948.

Hayashi, T. (1954). Effects of sodium glutamate on the nervous system. *Keio Journal of Medicine, 3*, 192–193.

Haynes, R. B., Devereaux, P. J., & Guyatt, G. H. (2002). Clinical expertise in the era of evidence-based medicine and patient choice. *American College of Physicians Journal Club, 136*, A11.

Heal, D. J., Smith, S. L., Gosden, J., & Nutt, D. J. (2013). Amphetamine, past and present – a pharmacological and clinical perspective. *Journal of Psychopharmacology, 27*, 479-496.

Health & Safety Alert #11-03-08. State of Ohio. Department of Mental Retardation and Developmental Disabilities. http://test.mr.state.oh.us/health/alerts/alert-11-3-08.pdf.

Healy, D. (2002). *The creation of psychopharmacology*. Cambridge, Massachusetts: Harvard University Press.

Healy, D. (2007). *Psychiatric drugs explained*. New York: Elsevier Churchill Livingstone.

Heiat, A., Gross, C., & Krumholz, H. (2002). Representation of the elderly, women, and minorities in heart failure clinical trials. *Archives of Internal Medicine, 162*, 1682-1688.

Heldt, J. (2017). *Memorable psychopharmacology*. (n.p.).

Hicks, R. W. *Summary of information submitted to MEDMARX in the year 2002: The quest for quality*. Washington, D.C.: The United States Pharmacopoeia.

Hilts, P. J. (2004). *Protecting America's health*. Chapel Hill, N.C.: The University of North Carolina Press.

Hogshire, J. (1999). *Pills-a-go-go*. Venice, California: Feral House.

Holst, C., Becker, U., Jorgensen, M. E., Gronbaek, M., & Tolstrup, J. S. (2017). Alcohol drinking patterns and risk of diabetes: A cohort study of 70,551 men and women from the general Danish population. *Diabetologia, 60*, 1941-1950.

Holzel, B. K., Carmody, J., Vangel, M., Congleton, C., Yerramsetti, S. M., Gard, T., & Lazar, S. W. (2011). Mindfulness practice leads to increases in regional brain gray matter density. *Psychiatry Research: Neuroimaging, 191*, 36-43.

How to prevent drug errors. (2010, July). *Consumer Reports on Health, 22*, 1+.

Howard, M. R. (2001). To improve the evidence of medicine: The 18th Century British origins of a critical approach. *Journal of the Royal Society of Medicine, 94*, 204-205.

Hsia, Y., Wong, A., Murphy, D., Simonoff, E., Buitelaar, J., & Wong, I. (2014). Psychopharmacological prescriptions for people with autism spectrum disorder (ASD): A multinational study. *Psychopharmacology, 231*(6), 999-1009.

Hunsley, J., & Di Giulio, G. (2002). Dodo bird, phoenix, or urban legend? The question of psychotherapy equivalence. *The Scientific Review of Mental Health Practice, 1*, 11-22.

Institute of Medicine. (1999). *To err is human: Building a safer health system*. Washington, D.C.: National Academy Press.

Institute of Medicine. (2001). *Crossing the quality chasm: A new health system for the 21st century*. Washington, DC: National Academy Press.

Institute of Medicine, Committee on the Assessment of Drug Safety Systems. (2011). *The future of drug safety: Promoting and protecting the health of the public*. Washington, D.C.: National Academic Press.

Interview with Merla Arnold. (2008, September). *The Tablet, Newsletter of Division 55 of the American Psychological Association, 9*(3), 8-10.

Iversen, L. (2000). Neurotransmitter transporters: Fruitful targets for CNS drug discovery. *Molecular Psychiatry, 5*, 357-362.

Iversen, L. (2006). Neurotransmitter transporters and their impact on the development of psychopharmacology. *British Journal of Pharmacology, 147 Supplement 1*, S82-88.

Jackson, C. R., Ruan, G., Aseem, F., Abey, J., Gamble, K., Stanwood, G., Palmiter, R. D., Iuvone, P., & McMahon, D. G. (2012). Retinal dopamine mediates multiple dimensions of light-adapted vision. *The Journal of Neuroscience, 32*, 9359-9368.

Jacob, S., & Spinler, S. A. (2006). Hyponatremia associated with Selective Serotonin-Reuptake Inhibitors in older adults. *Annals of Pharmacotherapy, 40*, 1618-1622.

Jenkins, T. A., Nguyen, J. C., Polglaze, K. E., & Bertrand, P. O. (2016). Influence of tryptophan and serotonin on mood and cognition with a possible role of the gut-brain axis. *Nutrients, 8*, 56.

Jensen, P. S., Arnold, L. E., Swanson, J., et al. (2007). Follow-up of the NIMH MTA study at 36 months after randomization. *Journal of American Academy of Child and Adolescent Psychiatry, 46*, 988-1001.

Jiang, X., Chen, W., Liu, X., Wang, Z., Liu, Y., Felder, R. A., Gildea, J. J., Jose, P. A., Qin, C., & Yang, Z. (2016). The synergistic roles of cholecystokinin B and dopamine D5 receptors on the regulation of renal sodium excretion. *PLoS One, 11*, 1-16.

Jones, P. B., Barnes, T. R., Davies, L., Dunn, G., Lloyd, H., Hayhurst, K. P., Murray, R. M., Markwick, A., & Lewis, S. W. (2006). Randomized controlled trial of the effect on quality of life of second- vs. first-generation antipsychotic drugs in schizophrenia: Cost utility of the latest antipsychotic drugs in schizophrenia study (CUtLASS 1). *Archives of General Psychiatry, 63*, 1079-1087.

Jordan, B. A., Cvejic, S., & Devi, L. A. (2000). Opioids and their complicated receptor complexes. *Neuropsychopharmacology, 23*(4 Suppl), S5–S18.

Julien, R. M. (2011). Psychopharmacology training in clinical psychology: A renewed call for action. *Journal of Clinical Psychology, 67*(4), 446-449.

Julien, R. M., Advokat, C. D., & Comaty, J. E. (2019). *A primer of drug action. A comprehensive guide to the actions, uses, and side effects of psychoactive drugs*. New York: Worth Publishers.

Kamienski, M., & Keogh, J. (2006). *Pharmacology demystified*. New York: McGraw Hill.

Kan, M. J., Lee, J. E., Wilson, J. G., Everhart, A. L., Brown, C. M., Hoofnagle, A. N., Jansen, M., Vitek, M. P., Gunn, M. D., & Colton, C. A. (2015). Arginine deprivation and immune suppression in a mouse model of Alzheimer's Disease. *The Journal of Neuroscience, 35*, 5969-5982.

Kane, J. M., Robinson, D. G., Schooler, N. R., Mueser, K. T., Penn, D. L., Rosenheck, R. A., Addington, J., Brunette, M. F., Correll, C. U., Estroff, S. E., Marcy, P., Robinson, J., Meyer-Kalos, P. S., Gottlieb, J. D., Glynn, S. M., Lynde, D. W., Pipes, R., Kurian, B. T., Miller, A. L., Azrin, S. T., Goldstein, A. B., Severe, J. B., Lin, H., Sint, K. J., John, M., & Heinssen, R. K. (2016). Comprehensive versus usual community care for first-episode psychosis: 2-Year outcomes from the NIMH RAISE early treatment program. *American Journal of Psychiatry, 173*, 362-372.

Kanjanarat, P., Winterstein, A. G., Johns, T. E., Hatton, R. C., Gonzaler-Rothi, R., & Segal, R. (2003). Nature of preventable adverse drug events in hospitals: A literature review. *American Journal of Health-System Pharmacy, 60*, 1750-1759.

Karet, G. B. (2019). How do drugs get named? *AMA Journal of Ethics*, 21, 686-696.

Keyte, L., & Dadson, M. (2014). Mediators and moderators to treatment response in large clinical trials. *Child & Adolescent Psychopharmacology News, 19*(4), 6-9.

Khan Academy. (2010, February 11). *Anatomy of a Neuron*. [Video]. YouTube. https://www.youtube.com/watch?v=ob5U8zPbAX4&t=97s

King, N. J., & Ollendick, T. H. (1998). Empirically validated treatments in clinical psychology. *Australian Psychologist, 33*, 89-95.

Kirk, S. A., Gomory, T., & Cohen, D. (2013). *Mad science*. New Brunswick, NJ: Transaction Publishers.

Kirsch, I. (2010). *The emperor's new drugs: Exploding the antidepressant myth*. New York: Basic Books.

Kirsch, I., & Sapirstein, G. (1998). Listening to Prozac but hearing placebo: A meta-analysis of antidepressant medication. *Prevention and Treatment, 1*, 2.

Klingenberg, M. (1958). Pigments of rat liver microsomes. *Archives of Biochemistry and Biophysics, 72*, 376-386.

Kondro, W., & Sibbald, B. (2004). Pharmaceutical industry – Drug company experts advised staff to withhold data about SSRI use in children. *Canadian Medical Association Journal, 170*, 783.

Konopka, L. M., & Zimmerman, E. M. (2014). Neurofeedback and psychopharmacology: Designing effective treatment based on cognitive and EEG effects of medications. In D. S. Cantor & J. R. Evans (Eds.), Clinical neurotherapy: Application of techniques for treatment (p. 55-84). Boston: Elsevier Academic Press.

Kopta, S. M., & Howard, K. I. (1994). Patterns of symptomatic recovery in psychotherapy. *Journal of Consulting and Clinical Psychology, 62*, 1009+.

Kreyenbuhl, J., Buchanan, R. W., Dickerson, F. B., & Dixon, L. B. (2010). The schizophrenia patient outcomes research team (PORT): Updated treatment recommendations 2009. *Schizophrenia Bulletin, 36*, 94-103.

Kushner, S. F., Khan, A., Lane, R., & Olson, W. H. (2006). Topiramate monotherapy in the management of acute mania: Results of four double-blind placebo-controlled trials. *Bipolar Disorders, 8*, 15-27.

Lam, D. H., Watkins, E. R., Hayward, P., Bright, J., Wright, K., Kerr, N., Parr-Davis, G., & Sham, P. (2003). A randomized controlled study of cognitive therapy for relapse prevention for bipolar affective disorder: Outcome of the First Year. *Archives of General Psychiatry, 60*(2), 145–152.

Lara, M. L. (2015). *The pharmacy in your kitchen: An overview of medical and medicinal foods.* Institute for Brain Potential Conference, Orlando, Florida.

Lasser, K. E., Allen, P. D., Woolhandler, S. J., Himmelstein, D. U., Wolfe, S. M., & Bor, D. H. (2002). Timing of new black box warnings and withdrawals for prescription medications. *Journal of American Medical Association, 287*, 2215-2220.

Lavoie, B., Lian, J. B., & Mawe, G. M. (2017). Regulation of bone metabolism by serotonin. *Advances in Experimental Medicine and Biology, 1033*, 35-46.

Lawyer, T. I., Jensen, J., & Welton, R. S. (2010). Serotonin syndrome in the deployed setting. *Military Medicine, 175,* 950-952.

Lazarou, J., Pomeranz, B. H., & Corey, P. N. (1998). Incidence of adverse drug reactions in hospitalized patients, a meta-analysis of prospective studies. *Journal of American Medical Association, 279*, 1200-1205.

Leary, T. (1988). *Politics of psychopharmacology.* Berkeley, CA: Ronin Publishing.

Lehman, A. (1999). Quality of care in mental health: The case of schizophrenia. *Health Affairs, 18*, 52-65.

Lehman, A. & Steinwachs, D. M., & Co-investigators. (1998). At issue: Translating research into practice: The schizophrenia Patient Outcomes Research Team (PORT) Treatment Recommendations. *Schizophrenia Bulletin, 24*, 1-10.

Lehman, A. F., Lieberman, J. A., Dixon, L. B., McGlashan, T. H., Miller, A. L., Perkins, D. O., & Kreyenbuhl, J. (2004). Practice guidelines for the treatment of patients with schizophrenia, 2nd edition. *American Journal of Psychiatry, 161*, 1-56.

Lehman, A. F., Steinwachs, D. M., & co-investigators. (1998a). Patterns of usual care for schizophrenia: Initial results from the schizophrenia patient outcomes research team (PORT) client survey. *Schizophrenia Bulletin, 24*, 11-20.

Lehman, A. F., Steinwachs, D. M., & co-investigators. (1998b). Translating research into practice: The schizophrenia patient outcomes research team (PORT) treatment recommendations. *Schizophrenia Bulletin, 24*, 1-10.

Lehman, A. F., Thompson, J. W., Dixon, L. B., & Scott, J. E. (1995). Schizophrenia: Treatment outcomes research—Editors' Introduction. *Schizophrenia Bulletin, 21*, 561-566.

Leichsenring, F., & Leibing, E. (2003). The effectiveness of psychodynamic therapy and cognitive behavior therapy in the treatment of personality disorders: A meta-analysis. *American Journal of Psychiatry, 160*, 1223-1232.

Leichsenring, F., & Rabung, S. (2008). Effectiveness of long-term psychodynamic psychotherapy: A meta-analysis. *Journal of American Medical Association, 300*, 1551-1565.

Lenzer, J. (2008, July). Medicine's magic bullets. *Discover, 29*, 46-52.

Leonhardt, D. (2006, February 22). Why doctors so often get it wrong. *The New York Times*.

Levi, M. I. (2007). *Basic notes in psychopharmacology*. New York: Radcliffe Publishing.

Levine, R., & Fink., M. (2006). The case against evidence-based principles in psychiatry. *Medical Hypotheses, 67*, 401-410.

Levinthal, C. F. (2012). *Drugs, behavior, & modern society*. Boston: Allyn & Bacon.

Levitin, D. J. (2020). *Successful aging: A neuroscientist explores the power and potential of our lives*. New York: Random House.

Li, G., Chen, Y., Hu, H., Liu, L., Hu, X., Wang, J., Shi, W., & Yin, D. (2012). Association between age-related decline of kidney function and plasma malondialdehyde. *Rejuvenation research, 15*, 257-264.

Li, R. M. (2019). Opioids. In K. Whalen (Ed.), *Pharmacology (7th Edition)* (pp. 180-193). Philadelphia: Wolters Kluwer.

Li, Y., Zhang, Y., Zhang, X. L., Feng, X. Y., Liu, C. Z., Zhang, X. N., Quan, Z. S., Yan, J. T., & Zhu, J. X. (2019). Dopamine promotes colonic mucus secretion through dopamine D5 receptor in rats. *American Journal of Physiology Cell Physiology, 316*, C393-403.

Lieberman, J. A. (2004). Metabolic changes associated with antipsychotic use. *The Primary Care Companion to the Journal of Clinical Psychiatry, 6*(2), 8-13.

Lieberman, J. A., Stroup, T. S., McEvoy, J. P., Swartz, M. S., Rosenheck, R. A., Perkins, D. O., Keefe, R. S., Davis, S. M., Davis, C. E., Lebowitz, B. D., Severe, J., & Hsiao, J. K. (2005). Effectiveness of antipsychotic drugs in patients with chronic schizophrenia. *New England Journal of Medicine, 353*, 1209-1223.

Liebowitz, M., & Tourian, K. A. (2010). Efficacy, safety, and tolerability of desvenlafaxine 50 mg/d for the treatment of major depressive disorder: A systematic review of clinical trials. *The Primary Care Companion to the Journal of Clinical Psychiatry, 12*, doi: 10.4088/PCC.09r00845blu: 10, 4088/PCC.09r00845blu

Liew, Z., Ritz, B., Rebordosa, C., Lee, P. C., & Olsen, J. (2014). Acetaminophen use during pregnancy, behavioral problems, and hyperkinetic disorders. *JAMA Pediatrics, 168*, 313-320.

Light, D. W. (2010). *The risks of prescription drugs*. New York: Columbia University Press.

Linehan, M. M., Armstrong, H. E., Suarez, A., Allmon, D., & Heard, H. L. (1991). Cognitive-behavioral treatment of chronically parasuicidal borderline patients. *Archives of General Psychiatry, 48*, 1060-1064.

Littlejohns, T. J., Neal, N. L., Bradbury, K. E., Heers, H., Allen, N. E., & Turney, B. W. (2020). Fluid intake and dietary factors and the risk of incident kidney stones in UK Biobank: A population-based prospective cohort study. *European Urology Focus, 6,* 752-761.

Littrell, J., & Ashford, J. B. (1995). Is it proper for psychologists to discuss medications with clients? *Professional Psychology: Research and Practice, 26*(3), 238-244.

Liu, C. Z., Zhang, X. L., Zhou, L., Wang, T., Quan, Z. S., Zhang, Y., Li, J., Li, G. W., Zheng, L. F., Li, L. S., & Zhu, J. X. (2018). Rasagiline, an inhibitor of MAO-B, decreases colonic motility through elevating colonic dopamine content. *Neurogastroenterology and Motility, 30*, e13390.

Liu, M., Ren, L., Zhong, X., Ding, Y., Liu, T., Liu, Z., Yang, X., Cui, L., Yang., L., Fan, Y., Liu, Y., & Zhang, Y. (2020). D2-like receptors mediate dopamine-inhibited insulin secretion via ion channels in rat pancreatic β-cells. *Frontiers in Endocrinology, 11*, 152.

Lofting, H. (1920). *The Story of Doctor Dolittle.* 100th Anniversary Edition. Orinda, CA: SeaWolf Press.

Lösel, F., & Schmucker, M. (2005). The effectiveness of treatment for sexual offenders: A comprehensive meta-analysis. *Journal of Experimental Criminology, 1*, 117-146.

Maes, M., Berk, M., Goehler, L., Song, C., Anderson, G., Galecki, P., & Leonard, B. (2012). Depression and sickness behavior are Janus-faced responses to shared inflammatory pathways. *BioMed Central Medicine, 10*, 66.

Maguire, D. R. & France, C. P. (2014). Impact of efficacy at the μ-opioid receptor on antinociceptive effects of combinations of μ-opioid receptor agonists and cannabinoid receptor agonists. *The Journal of Pharmacology and Experimental Therapeutics, 351,* 383-389.

Maidment, I. D., Haw, C., Stubbs, J., Fox, C., Katona, C., & Franklin, B. D. (2008). Medication errors in older people with mental health problems: A review. *International Journal of Geriatric Psychiatry, 23*, 564-573.

Maier, W., & Möller, H. J. (2010). Meta-analysis: A method to maximize the evidence from clinical studies? *European Archives of Psychiatry and Clinical Neuroscience, 260*, 17-23.

Makary, M. A., & Daniel, M. (2016). Medical error – The third leading cause of death in the US. *BMJ, 353*, i2139. Doi: 10.1136/bmj.i2139.

Malm, H., Sourander, A, Gissler, M., Gyllenberg, D., Hinkka-Yli-Salomaki, S., McKeague, I. W. Artama, M., & Brown, A. S. (2015). Pregnancy complications following prenatal exposure to SSRIs or maternal psychiatric disorders: Results from population-based National Register data. *American Journal of Psychiatry, 172*, 1224-1232.

Marcus, S. C., & Olfson, M. D. (2010). National trends in the treatment for depression from 1998 to 2007. *Archives of General Psychiatry, 67*, 1265-1273.

Marinho, V., Oliveira, T., Rocha, K., Ribeiro, J., Magalhaes, F., Bento, T., Pinto, G. R., Velasques, B., Ribeiro, P., Di Giorgio, L., Orsini, M., Gupta, D. S., Bittencourt, J., Bastos, V. H., & Teixeira, S. (2018). The dopaminergic system dynamic in the time perception: A review of the evidence. *The International Journal of Neuroscience, 128*, 262-282.

Maron, E., & Shlik, J. (2006). Serotonin function in panic disorder: Important, but why? *Neuropsychopharmacology: Official Publication of the American College of Neuropsychopharmacology, 31*, 1-11.

Martin, S. L., Power, A., Boyle, Y., Anderson, I. M., Silverdale, M. A., & Jones, A. K. P. (2017). 5-HT modulation of pain perception in humans. *Psychopharmacology, 234*, 2929-2939.

Martz, M. E., Trucco, E. M., Cope, L. M., Hardee, J. E., Jester, J. M., Zucker, R. A., & Heitzeg, M. M. (2016). Association of marijuana use with blunted nucleus accumbens response to reward anticipation. *JAMA Psychiatry, 73*, 838-844.

Mary Poppins. (1964). Burbank, CA: The Walt Disney Company.

Masand, P. S. (2000). Side effects of antipsychotics in the elderly. *Journal of Clinical Psychiatry, 61*, 43-49.

McCarter, K. D., Li, C., Jiang, Z., Lu, W., Smith, H. A., Xu, G., Mayhan, W. G., & Sun, H. (2017). Effect of low-dose alcohol consumption on inflammation following transient focal cerebral ischemia in rats. *Scientific Reports, 7*, 1-9.

McCarthy, J. (2015, April 1). *Mood-altering drug use highest in West Virginia, Lowest in Alaska.* Gallup. http://www.gallup.com/poll/182192/mood-altering-drug-highest-west-virginia-lowest-alaska.aspx.

McCoy, A. N., & Tan, S. Y. (2014). Otto Loewi (1873-1961): Dreamer and Nobel laureate. *Singapore Medical Journal, 55*, 3-4.

McDonnell, A. M., & Dang, C. H. (2013). Basic review of the Cytochrome P450 System. *Journal of the Advanced Practitioner in Oncology, 4*, 263-268.

McGill, M. R., & Jaeschke, H. (2013). Metabolism and disposition of acetaminophen: recent advances in relation to hepatotoxicity and diagnosis. *Pharmaceutical research, 30*, 2174–2187.

McGirr, A., & Berlim, M. T. (2014). A comment on Fond and colleagues' systematic review and meta-analysis of ketamine in the treatment of depressive disorders. *Psychopharmacology, 231*(19), 3907-3908.

McHugh, R. K., & Barlow, D. H. (2010). The dissemination and implementation of evidence-based psychological treatments, a review of current efforts. *American Psychologist, 65*, 73-84.

McKetin, R., Lubman, D., Baker, A. L., Dawe, S., & Ali, R. L. (2013). Dose-related psychotic symptoms in chronic methamphetamine users: Evidence from a prospective longitudinal study. *JAMA Psychiatry, 70*, 319-324.

Mechanic, D. (1998). Bringing science to medicine: The origins of evidence-based practice. *Health Affairs, 17*, 250-251.

Medicated Americans. (2008, February/March). *Scientific American Mind, 19*, 45.

Medication errors injure 1.5 million people and cost billions of dollars annually: report offers comprehensive strategies for reducing drug-related mistakes. (2006, July 20). The National Academies. http://www8.nationalacademies.org/onpinews/newsitem.aspx?RecordID=11623.

Medicines by design. (2006). Bethesda, MD: National Institute of Mental Health. https://www.nigms.nih.gov/education/Booklets/medicines-by-design/Pages/Home.aspx

Meier, B. (2018). *Pain killer: An empire of deceit and the origin of America's opioid epidemic*. New York: Random House.

Meldrum M. L. (2016). The ongoing opioid prescription epidemic: Historical context. *American Journal of Public Health, 106*(8), 1365–1366.

Mental health: A report of the Surgeon General. Washington, DC: Department of Health and Human Services, US Public Health Service, 2000.

Mental illness in America. (2008, February/March). *Scientific American Mind, 19*, 15.

Meyer, J. S., & Quenzer, L. F. (2019). *Psychopharmacology: Drugs, the brain, and behavior*. Sunderland, MA: Sinauer Associates, Inc.

Mitchell, D. C., Knight, C. A., Hockenberry, J., Teplansky, R., & Hartman, T. J. (2014). Beverage caffeine intakes in the U.S. *Food and Chemical Toxicology, 63*, 136-142.

Mitchell, J. M., Weinstein, D., Vega, T., & Kayser, A. S. (2018). Dopamine, time perception, and future time perspective. *Psychopharmacology, 235*, 2783-2793.

Mitte, K. (2005). A meta-analysis of the efficacy of psycho- and pharmacotherapy in panic disorder with and without agoraphobia. *Journal of Affective Disorders, 88*, 27-45.

Molina, B. S., Flory, K., Hinshaw, S. P., Greiner, A. R., Arnold, L. E., Swanson, J. M., Hechtman, L., Jensen, P. S., Vitiello, B., Hoza, B., Pelham, W. E., Elliott, G. R., Wells, K. C., Abikoff, H. B., Gibbons, R. D., Marcus, S., Conners, C. K., Epstein, J. N., Greenhill, L. L., ... Wigal, T. (2007). Delinquent behavior and emerging substance use in the MTA at 36-months: Prevalence, course, and treatment effects. *Journal of American Academy of Child and Adolescent Psychiatry, 46*, 1027-1039.

Möller, H. J. & Broich, K. (2009). Principle standards and problems regarding proof of efficacy in clinical psychopharmacology. *European Archives of Psychiatry and Clinical Neuroscience, 260*, 3-16.

Möller, H. J. & Maier, W. (2010). Evidence-based medicine in psychopharmacotherapy: Possibilities, problems and limitations. *European Archives of Psychiatry and Clinical Neuroscience, 260*, 25-39.

Moore, T. A., Buchanan, R. W., Buckley, P. F., Chiles, J. A., Conley, R. R., Crismon, M. L., Essock, S. M., Finnerty, M., Marder, S. R., Miller, D. D., McEvoy, J. P., Robinson, D. G., Schooler, N. R., Shon, S. P., Stroup, T. S., & Miller, A. L. (2007). The Texas medication algorithm project antipsychotic algorithm for schizophrenia: 2006 update. *Journal of Clinical Psychiatry, 68*, 1751-1762.

Moore, T. J. (1995). *Deadly medicine: Why tens of thousands of heart patients died in America's worst drug disaster*. New York: Simon & Schuster.

Moore, T. J. (1998). *Prescription for disaster: The hidden dangers in your medicine cabinet*. New York: Simon & Schuster.

Morrison, J. (1997). *When psychological problems mask medical disorders*. New York: The Guilford Press.

Moszczynska, A. (2016). Neurobiology and clinical manifestations of methamphetamine neurotoxicity. *The Psychiatric Times, 33*, 16-18.

Murray, C. J. L., & Frenk, J. (2011). Ranking 37th – Measuring the performance of the U.S. healthcare system. *The New England Journal of Medicine, 362*, 98-99.

Murray, E., Brouwer, S., McCutcheon, R., Harmer, C., Cowen, P., & McCabe, C. (2014). Opposing neural effects of naltrexone on food reward and aversion: Implications for the treatment of obesity. *Psychopharmacology, 231*(22), 4323-4335.

Murray, M., Hsia, Y., Glaser, K., Simonoff, E., Murphy, D., Asherson, P., Eklund, H., & Wong, I. (2014). Pharmacological treatments prescribed to people with autism spectrum disorder (ASD) in primary health care. *Psychopharmacology, 231*(6), 1011-1021.

Muse, M., & McGrath, R. E. (2010). Training comparison among three professions prescribing psychoactive medications: Psychiatric nurse practitioners, physicians, and pharmacologically trained psychologists. *Journal of Clinical Psychology, 66*, 96-103.

Muse, M., & Moore, B. A. (Eds). (2012). *Handbook of clinical psychopharmacology for psychologists.* New York: Wiley.

Nasi, G., Ahmed, T., Rasini, E., Fenoglio, D., Marino, F., Filaci, G., & Cosentino, M. (2019). Dopamine inhibits human CD8+ Treg function through D1-like dopaminergic receptors. *Journal of Neuroimmunology, 332*, 233-241.

Nathan, P. E., & Gorman, J. M. (Eds.). (1998). *A guide to treatments that work.* New York: Oxford University Press.

Neal, A., & Hogan, B. (2012). *Are your prescriptions killing you?* New York: Simon and Schuster.

Nestler, E. J., Hyman, S. W., & Malenka, R. C. (2009). *Molecular neuropharmacology: A foundation for clinical neuroscience.* New York: McGraw-Hill Medical.

Nestor, J. (2020). *Breath.* New York: Riverhead Books.

Neurology/Psychiatry. (2010, November). *Prescriber's Letter, 17*, 62.

Neurology/Psychiatry. (2011, December). *Prescriber's Letter, 18*(12), 70.

Nicol, N. (2007). Case study: An interdisciplinary approach to medication error reduction. *American Journal of Health-System Pharmacy, 64*, 17-20.

NIDA. 2020, April 8. What is the scope of marijuana use in the United States? Retrieved from https://www.drugabuse.gov/publications/research-reports/marijuana/what-scope-marijuana-use-in-united-states on 2020, July 22.

Nutt, D. J., & Blier, P. (2016). Neuroscience-based nomenclature (NbN) for Journal of Psychopharmacology. *Journal of Psychopharmacology, 30*, 413-415.

Nutt, D., Stahl, S., Blier, P., Drago, F., Zohar, J., & Wilson, S. (2017). Inverse agonists – what do they mean for psychiatry? *European Neuropsychopharmacology, 27*, 87-90.

Oikonomou, G., Altermatt, M., Zhang, R. W., Coughlin, G. M., Montz, C., Gradinaru, V., & Prober, D. A. (2019). The serotonergic raphe promote sleep in zebrafish and mice. *Neuron, 103*, 686-701.

Olfson, M., Blanco, C., Liu, L., Moreno, C., & Laje, G. (2006). National trends in the outpatient treatment of children and adolescents with antipsychotic drugs. *Archives of General Psychiatry, 63*, 679-685.

Osser, D. N. (2008). Cleaning up evidence-based psychopharmacology. *Psychopharm Review, 43*, 19-25.

Page, I. H. (1968). *Serotonin*. Chicago, IL: Yearbook Medical Publishers.

Parascandola, J. (2010). Abel, Takamine, and the isolation of epinephrine. *Journal of Allergy & Clinical Immunology, 125*, 514-517

Parker, M. (2005). False dichotomies: EBM, clinical freedom, and the art of the medicine. *Medical Humanities, 31*, 23-30.

Parsons, M. E., & Ganellin, C. R. (2006). Histamine and its receptors. *British Journal of Pharmacology, 147 Suppl1*, S127-S135.

Passani, M. B., Panula, P., & Lin, J. S. (2014). Histamine in the brain. *Frontiers in Systems Neuroscience, 8*, 64+.

Payne, K. S., Mazur, D. J., Hotaling, J. M., & Pastuszak, A. W. (2019). Cannabis and male fertility: A systematic review. *The Journal of urology, 202*(4), 674-681.

Perlmutter, A., Perlmutter, D., & Loberg, K. (2020). *Brain Wash*. New York: Little, Brown Spark.

Peterson, A. L. (2019). *Psych meds made simple*. Mental Health @ Home Books.

Petersen, M. (2008). *Our daily meds*. New York: Sarah Crichton Books.

Petrović, J., Pešić, V., & Lauschke, V. M. (2020). Frequencies of Clinically Important CYP2C19 and CYP2D6 Alleles are Graded across Europe. *European Journal of Human Genetics, 28*, 88-94.

Pinel, P. (1809). *Medico-Philosophical Treatise on Mental Alienation* (2nd edition). Entirely reworked and Extensively Expanded as translated by Hickish, G., Healy, D., Charland L. C. (2008), Chichester, West Sussex: John Wiley and Sons.

Pinel, P., Charland, L., Haely D., Hickish, G., & Weiner, D. (2008). *Medico-philosophical treatise on mental alienation*. Hoboken, N.J.: Wiley-Blackwell.

Plener, P. L., & Schulze, U. M. (2014). Pharmacological treatment of non-suicidal self-injury and eating disorders. In L. Claes & J. J. Muehlenkamp (Eds.), *Non-suicidal self-injury in eating disorders: Advancements in etiology and treatment* (pp. 197-214). New York, NY, US: Springer-Verlag Publishing.

Powers, M. B., & Emmelkamp, P. M. (2008). Virtual reality exposure therapy for anxiety disorders: A meta-analysis. *Journal of Anxiety Disorders, 22*, 561-569.

Practice guidelines regarding psychologists' involvement in pharmacological issues. (2011). *American Psychologist, 66*(9), 835-849.

Prauss, K., Varatharajan, R., Joseph, K., & Moser, A. (2014). Transmitter self-regulation by extracellular glutamate in fresh human cortical slices. *Journal of Neural Transmission, 121*, 1321-1327.

President's New Freedom Commission on Mental Health. (2003). *Achieving the promise: Transforming mental health care in America. Final Report.* (DHHS PUB. No. SMA-03-3832). Rockville, MD: U.S. Government Printing Office.

Preston, J., & Johnson, J. (2016). *Clinical psychopharmacology made ridiculously simple.* Miami, Florida: Medmaster.

Preston, J. D., O'Neal, J. H., & Talaga, M. C. (2008). *Handbook of clinical psychopharmacology for therapists.* Oakland, CA: New Harbinger Publications, Inc.

Radley, D. C., Finkelstein, S. N., & Stafford, R. S. (2006). Off-label prescribing among office-based physicians. *Archives of Internal Medicine, 166*, 1021-1026.

Ragusea, S. (Winter, 2011). Defining competence. *Florida Psychological Association, 62*, 18.

Ramey, K., Ma, J. D., Best, B. M., Atayee, R. S., & Morello, C. M. (2014). Variability in metabolism of imipramine and desipramine using urinary excretion data. *Journal of Analytical Toxicology, 38*, 368–374.

Ramos-Quiroga, J. A., Corominas-Roso, M., Palomar, G., Gomez-Barros, N., Ribases, M., Sanchez-Mora, C., Bosch, R., Nogueira, M., Corrales, M., Valero, S., & Casas, M. (2014). Changes in the serum levels of brain-derived neurotrophic factor in adults with attention deficit hyperactivity disorder after treatment with atomoxetine. *Psychopharmacology, 231*(7), 1389-1395.

Rao, Y. (2019). The first hormone: Adrenaline. *Trends in Endocrinology and Metabolism, 30*, 331-334.

Reeves, R. R., Ladner, M. E., Hart, R. H., & Burke, R. S. (2007). Nocebo effects with antidepressant clinical drug trial placebos. *General Hospital Psychiatry, 29*, 275-277.

Reiss, D. M. (March 1, 2005). Wisdom-based treatment. *Psychiatric Times, 22*, 95.

Renda, T. G. (2000). Vittorio Erspamer: A true pioneer in the field of bioactive peptides. *Peptides, 21*, 1585-1586.

Ridker, P. M., Danielson, E., Fonseca, F. A., Genest, J., Gotto, A. M., Jr, Kastelein, J. J., Koenig, W., Libby, P., Lorenzatti, A. J., MacFadyen, J. G., Nordestgaard, B. G., Shepherd, J., Willerson, J. T., & Glynn, R. J. (2008). Rosuvastatin to prevent vascular events in men and women with elevated c-reactive protein. *New England Journal of Medicine, 359*, 2195-2207.

Roberts, E., & Frankel, S. (1950). Gamma-aminobutyric acid in brain: Its formation from glutamic acid. *Journal of Biological Chemistry, 187*, 55-63.

Rodwin, M. (2001). The politics of evidence-based medicine. *Journal of Health Politics, Policy, and Law, 26*, 439-446.

Roelofs, K. (2017). Freeze for action: Neurobiological mechanisms in animal and human freezing. *Philosophical transactions of the Royal Society of London. Series B, Biological Sciences, 372*, 20160206.

Rogers, W. A. (2004). Evidence-based medicine and justice: A framework for looking at the impact of EBM upon vulnerable or disadvantaged groups. *Journal of Medical Ethics, 30*, 141-145.

Rosenheck, R. (2005). The growth of psychopharmacology in the 1990s: Evidence-based practice or irrational exuberance. *International Journal of Law and Psychiatry, 28*, 467-483.

Rosenzweig, S. (1936). Some implicit common factors in diverse methods of psychotherapy. *American Journal of Orthopsychiatry, 6*, 412-415.

Ross, H. R., Napier, I., & Connor, M. (2008). Inhibition of recombinant human T-type calcium channels by Delta-9-tetrahydrocannabinol and cannabidiol. *The Journal of Biological Chemistry, 283*, 16124-16134.

Rowling, J. K. (2003). *Harry Potter and the order of the phoenix*. New York: Scholastic.

Sachdev, P. (1995). The development of the concept of akathisia: A historical overview. *Schizophrenia Research, 16*, 33-45.

Sackett, D. L. (1997). Evidence-based medicine. *Seminars in Perinatology, 21*, 3-5.

Sackett, D. L., Rosenberg, W., Muir Gray, J. A., Haynes, R. B., & Richardson, W. S. (1996). Evidence based medicine: What it is and what it isn't. *British Medical Journal, 312*, 71-72.

Sadock, B. J., & Sadock, V. A. (2003). *Kaplan & Sadock's synopsis of psychiatry: Behavioral sciences/clinical psychiatry (9th edition)*. Philadelphia: Lippincott, Williams, & Wilkins.

Saey, T. H. (2010). Let there be light: New technology illuminates neuronal conversations in the brain. *Science News, 177*(3), 18-21.

Saks, O. (1973). *Awakenings*. New York: Doubleday.

Salzman, C., Glick, I., & Keshavan, M. S. (2010). The 7 sins of psychopharmacology. *Journal of Clinical Psychopharmacology, 30*, 653-655.

Sammons, M. T., & Schmidt, N. B. (Eds.). (2001). *Combined treatments for mental disorders: A guide to psychological and pharmacological interventions*. Washington, D.C.: American Psychological Association.

Sansone, R. A., & Sansone, L. A. (2010). SSRI-induced indifference. *Psychiatry, 7*, 14-18.

Sansone, R. A., & Sansone, L. A. (2012). SSRIs: Bad to the bone? *Innovations in Clinical Neuroscience, 9*, 42-49.

Sarteschi, C. M. (2014). Randomized controlled trials of psychopharmacological interventions of children and adolescents with conduct disorder: A descriptive analysis. *Journal of Evidence-Based Social Work, 11*(4), 350-359.

Saunders, J.C., Kline, N.S., Vaisberg, M., Munster, A.J. & Bailey, S. D'A. (1959). Psychic energizers. In J. Masserman (Ed.), *Biological Psychiatry*: The Proceedings of the scientific sessions of the society of biological psychiatry (pp. 306-310). New York: Grune & Stratton.

Sawyer, A., Lake, J. K., Lunsky, Y., Liu, S., & Desarkar, P. (2014). Psychopharmacological treatment of challenging behaviors in adults with autism and intellectual disabilities: A systematic review. *Research in Autism Spectrum Disorders, 8*(7), 803-813.

Scarante, F. F., Vila-Verde, C., Detoni, V. L., Ferreira-Junior, N. C., Guimaraes, F. S., & Campos, A. C. (2017). Cannabinoid modulation of the stressed hippocampus. *Frontiers in Molecular Neuroscience, 10*, 411+.

Schatzberg, A. F., & DeBattista, C. (2019). *Schatzberg's manual of clinical psychopharmacology (9th Edition)*. Washington, D.C.: American Psychiatric Association.

Schiff, G. D., Galanter, W. L., Duhig, J., Lodolce, A. E., Koronkowski, M. J., & Lambert, B. L. (2011). Principles of conservative prescribing. *Archives of Internal Medicine, 171*, 1433-1439.

Schildkraut, J. J. (1965). The catecholamine hypothesis of affective disorders: A review of supporting evidence. *American Journal of Psychiatry, 122*, 509-522.

Schmidt, N. B., Richey, J. A., Zvolensky, M. J., & Maner, J. K. (2008). Exploring human freeze responses to a threat stressor. *Journal of Behavior Therapy and Experimental Psychiatry, 39*, 292-304.

Schneider, L. S., Dagerman, K. S., & Insel, P. (2005). Risk of death with atypical antipsychotic drug treatment for dementia: Meta-analysis of randomized placebo-controlled trials. *Journal of American Medical Association, 294*, 1934-1943.

Schoenwald, S. K., Sheidow, A. J., Letourneau, E. J., & Liao, J. G. (2003). Transportability of multisystemic therapy: Evidence for multilevel influences. *Mental Health Services Research, 5*, 223-239.

Scott, J., Gade, G., McKenzie, M., & Venohr, I. (1998). Cooperative health care clinics: A group approach to individual care. *Geriatrics, 53*, 68+.

Segal, Z. V., Bieling, P., Young, T., MacQueen, G., Cooke, R., Martin, L., Bloch, R., & Levitan, R. D. (2010). Antidepressant monotherapy vs. sequential pharmacotherapy and mindfulness-based cognitive therapy, or placebo, for relapse prophylaxis in recurrent depression. *Archives of General Psychiatry, 67*, 1256-1264.

Selvaraj, S., Kuman, N., Elakiya, M. M., Saraswathy, C., Balaji, D., & Mohan, S. (2010). Evidence-based medicine, a new approach to teach medicine: A basic review for beginners. *Biology & Medicine, 2*, 1-5.

Seminar Advertisement, "Malnourished Minds: Integrative Medicine for Mood Disorders." (2011, October). Advertising Brochure, Cross Country Education, www.CrossCountryEducation.com.

Sena, S. F., Kazimi, S., & Wu, A. H. (2002). False-positive phencyclidine immunoassay results caused by venlafaxine and O-desmethylvenlafaxine. *Clinical Chemistry, 48*, 676-677.

Sepkowitz, K. A. (2013). Energy drinks and caffeine-related adverse effects. *Journal of American Medical Association, 309*, 243-244.

Settle, E. C., Jr. (1998). Bupropion sustained release: Side effect profile. *Journal of Clinical Psychiatry, 59*, 32-36.

Shah, A. (2007). Drugs you might stop: A practical approach to medication debridement. *Family Practice Recertification, 29*, 45-51.

Shakespeare, W. (1594-1596). *Romeo and Juliet*.

Shakespeare, W. (1599). *The Tragedy of Julius Caesar*.

Shedler, J. (2010). The efficacy of psychodynamic psychotherapy. *American Psychologist, 65*, 98-109.

Sherman, R., & Hickner, J. (2007). Academic physicians use placebos in clinical practice and believe in the mind-body connection. *Journal of General Internal Medicine, 23*, 7-10.

Shuval, K., Linn, S., Brezis, M., Shadmi, E., Green, M. L., & Reis, S. (2010). Association between primary care physicians' evidence-based medicine knowledge and quality of care. *International Journal for Quality in Health Care, 22*, 16-23.

Sian, J., Youdim, M. B. H., Riederer. P., & Gerlach, M. (1999). Biochemical anatomy of the basal ganglia and associated neural systems. In G. J. Siegel, B. W. Agranoff, & R. W. Albers (Eds.). Basic neurochemistry: Molecular, cellular and medical aspects. 6th edition. Philadelphia: Lippincott-Raven.

Sifferlin, A. (2017, January 16). How Botox became the drug that's treating everything. *Time*, p. 38.

Sinclair, D. C., Purves-Tyson, T. D., Allen, K. M., & Weickert, C. S. (2014). Impacts of stress and sex hormones on dopamine neurotransmission in the adolescent brain. P*sychopharmacology, 231*(8), 1581-1599.

Siversten, B., Omvik, S., Pallesen, S., Bjorvatn, B., Havik, O. E., Kvale, G., Nielsen, G. H., & Nordhus, I. H. (2006). Cognitive behavioral therapy vs. zopiclone for treatment of chronic primary insomnia in older adults: A randomized controlled trial. *Journal of American Medical Association, 295*, 2851+.

Skinner, M. D., Lahmek, P., Pham, H., & Aubin, H. (2014). Disulfiram efficacy in the treatment of alcohol dependence: A meta-analysis. *Plos ONE, 9*(2), 1-15.

Slater, L. (2018). *Blue dreams*. New York: Little, Brown and Company.

Smith, B. L. (2012, June). Inappropriate prescribing. *Monitor on Psychology, 43*(6), 36+.

Smith, J. (2014). Review of anxiety disorders: A guide for integrating psychopharmacology and psychotherapy. *Child and Adolescent Mental Health, 19*(1), 79-80.

Smith, M. C. (1985). *Small comfort: A history of the minor tranquilizers*. New York: Praeger Scientific.

Smith, M. L., Glass, G. V., & Miller, T. L. (1980). *The benefits of psychotherapy*. Baltimore: Johns Hopkins University Press.

Sneed, J. R., Reinlieb, M. E., Rutherford, B. R., Miyazaki, M., Fitzsimons, L., Turret, N., Pelton, G. H., Devanand, D. P., Sackeim, H. A., & Roose, S. P. (2014). Antidepressant treatment of melancholia in older adults. *The American Journal of Geriatric Psychiatry, 22*(1), 46-52.

Sontineni, S. P., Chaudhary, S., Sontineni, V., & Lanspa, S. J. (2009). Cannabinoid hyperemesis syndrome: Clinical diagnosis of an under-recognized manifestation of chronic cannabis abuse. *World Journal of Gastroenterology, 15*, 1264-1266.

Spiegel, R. (2003). *Psychopharmacology: An introduction*. West Sussex, England: Wiley.

Spiering, M. J. (2018). The discovery of GABA in the brain. *The Journal of Biological Chemistry, 293*, 19159-19160.

Srivastava, A., & Coffey, B. (2014). Emotional dysregulation in a child with attention-deficit/hyperactivity disorder and anxiety: Psychopharmacological strategies. *Journal of Child and Adolescent Psychopharmacology, 24*(10), 590-593.

Stack, E. (2006). Physiotherapy: The ultimate placebo. *Physiology Research, 11*, 127-128.

Stahl, S. M. (2008). *Stahl's essential psychopharmacology: Neuroscientific basis and practical applications*. New York: Cambridge University Press.

Stahl, S. M. (2017). *The prescriber's guide*. New York: Cambridge University Press.

Staying safe in the hospital. (2010). *Consumer Reports On Health, 22*(11), 1.

Steckler, T., Spooren, W., & Murphy, D. (2014). Autism spectrum disorders—an emerging area in psychopharmacology. *Psychopharmacology, 231*(6), 977-978.

Sternbach, H. (1991). The serotonin syndrome. *American Journal of Psychiatry, 148*, 705-713.

Stoppler, M. C. *The most common medication errors*. MedicineNet. http://www.medicinenet.com/script/main/art.asp?articlekey=55234&pf=3&page=1.

Straus, S. E., & McAlister, F. A. (2000). Evidence-based medicine: A commentary on common criticisms. *Canadian Medical Association Journal, 163*, 837-841.

Swanson, J. M., Arnold, L. E., Kraemer, H., Hechtman L., Molina, B., Hinshaw, S., Vitiello, B., Jensen, P., Steinhoff, K., Lerner, M., Greenhill, L., Abikoff, H., Wells, K., Epstein, J., Elliott, G., Newcorn, J., Hoza, B., & Wigal, T. (2008a). Evidence, interpretation, and qualification from multiple reports of long-term outcomes in the Multimodal Treatment Study of Children with ADHD (MTA), part I: Executive summary. *Journal of Attention Disorders, 12*, 4-14.

Swanson, J. M., Arnold, L. E., Kraemer, H., Hechtman, L., Molina, B., Hinshaw, S., Vitiello, B., Jensen, P., Steinhoff, K., Lerner, M., Greenhill, L., Abikoff, H., Wells, K., Epstein, J., Elliott, G., Newcorn, J., Hoza, B., & Wigal, T. (2008b). Evidence, interpretation, and qualification from multiple reports of long-term outcomes in the Multimodal Treatment Study of Children with ADHD (MTA), part II: Supporting details. *Journal of Attention Disorders, 12*, 15-43.

Takagaki, G. (1996). The dawn of excitatory amino acid research in Japan. The pioneering work by Professor Takashi Hayashi. *Neurochemistry International, 29,* 225-229.

Tansey, E. M. (2006). Henry Dale and the discovery of acetylcholine. *Comptes Rendus-Biologies, 329,* 419-425.

Taylor, D., Paton, C., & Kapur, S. (2012). *The Maudsley prescribing guidelines in psychiatry (11ᵗʰ Edition)*. Hoboken, NJ: Wiley-Blackwell.

Taylor, S. E. (2009). *Health psychology*. New York: McGraw-Hill.

The Rolling Stones. (1966). "Mother's Little Helper." *Aftermath*. Decca – DL25 260, Format: Vinyl 7.

The Schizophrenia Patient Outcomes Research Team (PORT) Treatment Recommendations. At issue: Translating Research into Practice. http://www.ahrq.gov/clinic/schzrec.htm.

Therapy vs. drugs for depression. (2011). *Consumer Reports on Health, 23*, 6.

Thornley, B., & Adams, C. (2000). Content and quality of 2000-controlled trials in schizophrenia over 50 years. *British Medical Journal, 317*, 1181-1184.

Timmermans, S., & Mauck, A. (2005). The promises and pitfalls of evidence-based medicine. *Health Affairs, 24*, 18-28.

Tobin, J. (2008). Counterpoint: Evidence-based medicine lacks a sound scientific base. *Chest, 133*, 1071-1074.

Too much medication. (2012, March). *Consumer Reports on Health, 24*, 1+.

Towner, B. (2009, October). The 50 most prescribed drugs. *AARP Bulletin, 50,* 39.

Tripathi, R., Rao, R., Dhawan, A., Jain, R., & Sinha, S. (2020). Opioids and sleep – a review of literature. *Sleep Medicine, 67*, 269-275.

Turner, E. H., Matthews, A. M., Linardatos, E., Tell, R. A., & Rosenthal, R. (2008). Selective publication of antidepressant trials and its influence on apparent efficacy. *New England Journal of Medicine, 358*, 252-260.

Twarog, B. M. (1954). Responses of a molluscan smooth muscle to acetylcholine and 5-hydroxytryptamine. *Journal of Cellular and Comparative Physiology, 44*, 141-163.

Understanding and managing the pieces of major depressive disorder. (2009). Carlsbad, CA: Neuroscience Education Institute.

Underwood, M. D., Kassir, S. A., Bakalian, M. J., Galfalvy, H., Dwork, A. J., Mann, J. J., & Arango, V. (2018). Serotonin receptors and suicide, major depression alcohol use disorder and reported early life adversity. *Translational Psychiatry, 8*, 1-15.

United States v. Forty Barrels and Twenty Kegs of Coca Cola, 241 U.S. 265 (1916).

US Equal Opportunity Employment Commission. *Genetic Information Discrimination*. https://www.eeoc.gov/laws/types/genetic.cfm

Users' guides to the medical literature. (1993). *Journal of American Medical Association, 270*, 2096-2097.

Van Bockstaele, E. J., & Valentino, R. J. (2013). Neuropeptide regulation of the locus coeruleus and opiate-induced plasticity of stress responses. *Advances in Pharmacology, 68*, 405-420.

Van Putten, T. (1975). The many faces of akathisia. *Comprehensive Psychiatry, 16*, 43-47.

Van Ryswyk, E., & Antic, N. A. (2016). Opioids and sleep-disordered breathing. *Chest, 150*, 934-944.

Vasa, R. A., Carroll, L. M., Nozzolillo, A. A., Mahajan, R., Mazurek, M. O., Bennett, A. E., Wink, L. K., & Bernal, M. P. (2014). A systematic review of treatments for anxiety in youth with autism spectrum disorders. *Journal of Autism and Developmental Disorders, 44*(12), 3215-3229.

Vázquez, G. H. (2014). The impact of psychopharmacology on contemporary clinical psychiatry. *Canadian Journal of Psychiatry, 59*(8), 412-416.

Verachai, V., Rukngan, W., Chawanakrasaesin, K., Nilaban, S., Suwanmajo, S., Thanateerabunjong, R., Kaewkungwal, J., & Kalayasiri, R. (2014). Treatment of methamphetamine-induced psychosis: A double-blind randomized controlled trial comparing haloperidol and quetiapine. *Psychopharmacology, 231*(16), 3099-3108.

Vuong, C., Van Uum, S. H., O'Dell, L. E., Lutfy, K., & Friedman, T. C. (2010). The effects of opioids and opioid analogs on animal and human endocrine systems. *Endocrine Reviews, 31*, 98-132.

Wakefield, M., & DeLeon, P. H. (2000, June). To err is human: An Institute of Medicine report. *Professional Psychology: Research & Practice, 31*, 243-244.

Walker, O. S., Holloway, A. C., & Raha, S. (2019). The role of the endocannabinoid system in female reproductive tissues. *Journal of Ovarian Research, 12*(1), 1–10.

Wampold, B. E., Lichtenberg, J. W., & Waehler, C. A. (2002). Principles of empirically supported interventions in counseling psychology. *The Counseling Psychologist, 30*, 197-217.

Wang, J., Duan, X., Li, B., & Jiang, X. (2017). Alcohol consumption and risk of gallstone disease: A meta-analysis. *European Journal of Gastroenterology & Hepatology, 29*, e19-e28.

Wang, L., Wu, C., Lee, S., & Tsai, Y. (2014). Salivary neurosteroid levels and behavioural profiles of children with attention-deficit/hyperactivity disorder during six months of methylphenidate treatment. *Journal of Child & Adolescent Psychopharmacology, 24*(6), 336-340.

Wang, P., Pradhan, K., Zhong, X. B., & Ma, X. (2016). Isoniazid metabolism and hepatotoxicity. *Acta Pharmaceutica Sinica. B*, 384–392.

Wang, P. S., Berglund, P., & Kessler, R. C. (2000). Recent care of common mental disorders in the United States: Prevalence and conformance with evidence-based recommendations. *Journal of General Internal Medicine, 15*, 284-292.

Waning, B., & Montagne, M. (2001). *Pharmacoepidemiology: principles and practice.* New York: McGraw-Hill.

Watkins, J. C. & Jane, D. E. (2006). The glutamate story. *British Journal of Pharmacology, 147*, S100-S108.

Wegmann, J. (2015). *Psychopharmacology: Straight talk on mental health medications.* Eau Claire, WI: PESI Publishing & Media.

Weissman, D. E., & Haddox, J. D. (1989). Opioid pseudoaddiction — an iatrogenic syndrome. *Pain, 36*, 363-366.

Weisz, J. R., Jensen-Doss, A., & Hawley, K. M. (2006). Evidence-based youth psychotherapies versus usual clinical care: A meta-analysis of direct comparisons. *American Psychologist, 61*, 671-689.

Weisz, J. R., Weiss, B., Alicke, M. D., & Klotz, M. L. (1987). Effectiveness of psychotherapy with children and adolescents: A meta-analysis for clinicians. *Journal of Consulting and Clinical Psychology, 55*, 542-549.

Weisz, J. R., Weiss, B., Han, S., Granger, D. A., & Morton, T. (1995). Effects of psychotherapy with children and adolescents revisited: A meta-analysis of treatment outcome studies. *Psychological Bulletin, 117*, 450-468.

Wenner, M. (2010). Are antidepressants safe for pregnant women? *Scientific American Mind, 19*, 8.

Westmacott, R., & Hunsley, J. (2007). Weighing the evidence for psychotherapy equivalence: Implications for research and practice. *The Behavior Analyst Today, 8*, 210-225.

Whalen, K., Radhakrishnan, R., & Field, C. (Eds.). (2019). *Lippincott Illustrated Reviews Pharmacology (6th Edition)*. Alphenaan den Rijn, The Netherlands: Wolters Kluwer Publisher.

Whitaker, R. (2002). *Mad in America*. New York: Basic Books.

Whitaker, R. (2010). *Anatomy of an epidemic*. New York: Crown.

Whitaker, R. (2011). *Now antidepressant-induced chronic depression has a name: Tardive Dysphoria*. Psychology Today. http://www.psychologytoday.com/print/68229.

Whitaker, R., & Cosgrove, L. (2015). *Psychiatry under the Influence*. New York: Palgrave MacMillan

Whitaker-Azmitia, P. M. (1999). The discovery of serotonin and its role in neuroscience. *Neuropsychopharmacology, 21*, 2-8.

Whittal, M. L., Agras, W. S., & Gould, R. A. (1999). Bulimia nervosa: A meta-analysis of psychosocial and pharmacological treatments. *Behavior Therapy, 30*, 117-135.

Williams, R., & Stone, J. (2012, September 17). The Nazis at the heart of the worst drug scandal of all time. *Newsweek*, 34-39.

Wilson, G. T., & Fairburn, C. G. (1998). Treatments for eating disorders. In P. E. Nathan & J. M. Gorman (Eds.), *A guide to treatments that work*. New York: Oxford University Press.

Winkelman, N. W. (1954). Chlorpromazine in the treatment of neuropsychiatric disorders. *Journal of American Medical Association, 155*, 18-21.

Wise, M. E. J. (2001). Citalopram-induced bruxism. *The British Journal of Psychiatry, 178*, 182.

Wooley, D. W. (1962). *The biochemical basis of psychoses or the serotonin hypothesis about mental diseases*. New York: John Wiley & Sons, Inc.

Wu, K. M. (2009). A new classification of prodrugs: Regulatory perspectives. *Pharmaceuticals, 2*, 77-81.

Wu, M., Sahbaie, P., Zheng, M., Lobato, R., Boison, D., Clark, J. D., & Peltz, G. (2013). Opiate-induced changes in brain adenosine levels and narcotic drug responses. *Neuroscience, 228*, 235-242.

Yano, J. M., Yu, K., Donaldson, G. P., Shastri, G. G., Ann, P., Ma, L., Nagler, C. R., Ismagilov, R. F., Mazmanian, S. K., & Hsiao, E. (2015). Indigenous bacteria from the gut microbiota regulate host serotonin biosynthesis. *Cell, 161*, 264-276.

Zhang, C., Li, Z., Wu, Z., Chen, J., Wang, Z., Peng, D., Hong, W., Yuan, C., Wang, Z., Yu, S., Xu, Y., Xu, L., Xiao, Z., & Fang, Y. (2014). A study of N-methyl-D-aspartate receptor gene (GRIN2B) variants as predictors of treatment-resistant major depression. *Psychopharmacology, 231*(4), 685-693.

Zimmer, B., Chiodo, K., & Roberts, D. (2014). Reduction of the reinforcing effectiveness of cocaine by continuous d-amphetamine treatment in rats: Importance of active self-administration during treatment period. *Psychopharmacology, 231*(5), 949-954.

Zohar, J., Stahl, S., Moller, H., Blier, P., Kupfer, D., Yamawaki, S., Uchida, H., Spedding, M., Goodwin, G., & Nutt, D. (2015). A review of the current nomenclature for psychotropic agents and an introduction to the Neuroscience-based Nomenclature. *European Neuropsychopharmacology, 25*, 2318-2325.

414